104 Advances in Polymer Science

Polyelectrolytes Hydrogels Chromatographic Materials

With contributions by
S. A. Dubrovskii, A. E. Ivanov, K. S. Kazanskii,
N. P. Kuznetsova, V. V. Saburov,
G. V. Samsonov, V. P. Zubov

With 75 Figures and 33 Tables

Springer-Verlag
Berlin Heidelberg NewYork
London Paris Tokyo
Hong Kong Barcelona
Budapest

ISBN-3-540-55109-3 Springer-Verlag Berlin Heidelberg New York
ISBN-0-387-55109-3 Springer-Verlag New York Berlin Heidelberg

Library of Congress Catalog Card Number 61-642

Typesetting: Th. Müntzer, Bad Langensalza; Printing: Heenemann, Berlin;
Bookbinding: Lüderitz & Bauer, Berlin
02/3020-5 4 3 2 1 0 – Printed on acid-free paper

Editors

Table of Contents

Crosslinked Polyelectrolytes in Biology

Georgy V. Samsonov, Nina P. Kuznetsova
Institute of Macromolecular Compounds, Academy of Sciences,
S.-Petersburg, Bolshoi prospect 31, 199004, USSR

This review deals with the structure and properties of highly permeable crosslinked polyelectrolytes synthesized mainly by using acrylic acids and various polyvinyl crosslinking agents. Special attention was directed to macro- and heteroreticular polyelectrolytes with high inhomogeneity. The mobility of parts of polymer chains located between chemical and physical junctions was demonstrated. The biological aspect of these systems was analyzed on the basis of thermodynamic selectivity and kinetic relationships in the interaction between crosslinked polyelectrolytes and biologically active molecules and ions.

Special emphasis was put on the properties and application of the microdisperse forms of crosslinked polyelectrolytes. The possibilities of using highly permeable crosslinked polyelectrolytes for immobilization of proteins (and enzymes) and new approaches to the use of crosslinked polyelectrolytes biosorbents in preparative chromatography were also considered.

Advances in Polymer Science Vol. 104

List of Abreviations

CP	crosslinked polyelectrolyte
AA	acrylic acid
MA	methacrylic acid
PAA	polyacrylic acid
PMA	polymethacrylic acid
DVB	divinylbenzene
AA-DVB	acrylic acid-divinylbenzene copolymer
St-DVB	styrene-divinylbenzene copolymer
EDMA	*N,N'*-ethylene dimethacrylamide
HHTT	hexahydro-1,3,5-triacryloyltriazine
DMEG	ethylene glycol dimethacrylate
Ac A	acetic acid
gr.	grain
AA-EDMA	acrylic acid-ethylene dimethacrylamide copolymer
MA-EDMA	methacrylic acid-ethylene dimethacrylamide copolymer
MA-DMEG	methacrylic acid-ethylene glycol dimethacrylate copolymer
MA-HHTT	(Biocarb-T) – methacrylic acid-hexahydro-1,3,5-triacryloyltriazine copolymer
BAS	biologically active substance
Hb	haemoglobin
SA	serum albumin
Ls	lysozyme
P	protein
Im Hb	immobilized Hb in microdispersion of CP
KU-1	sulfophenol condensed with formaldehyde
KU-2	sulfonated copolymer of styrene and divinylbenzene
KU-23	macroporous sulfonated copolymer of styrene and divinylbenzene

SDV-T	telagenated product of sulfonated copolymer of styrene and divinylbenzene
KRC	sulfonated copolymers of styrene and *p*-divinylbenzene (4,5 mol% of crosslinking agent)
SBS	sulfonated copolymer of styrene and butadiene
Dowex-50	sulfonated copolymer of styrene and divinylbenzene
SDMDMA	sulfonated copolymer of methacrylamide and *N,N'*-decamethylene dimethacrylamide
SHMDMA	sulfonated copolymer of methacrylamide and *N,N'*-hexamethylene dimethacrylamide
SEDMA	sulfonated copolymer of methacrylamide and *N,N'*-ethylene dimetacrylamide
CS-KU-2 (KU-23)	KU-2 (KU-23) immobilized in cellulose grains
KB-2	copolymer of acrylic acid and divinylbenzene
KB-4	copolymer of methacrylic acid and divinylbenzene
KMDM-6	copolymer of methacrylic acid and *N,N'*-hexamethylene dimethacrylamide
KFA	condensation product of phenoxyacetic acid with formaldehyde
AV-16	condensation product of pyridine and polyethylene polyamines with epichlorohydrin
AV-17	aminated chloromethylated copolymer of styrene and divinylbenzene
FAV-16 FAV-17	microdisperse grains of AV-16 or AV-17
CS-AV-17	AV-17 immobilized in cellulose grains
ARA	aminated chloromethylated copolymer of styrene and *p*-divinylbenzene
FAF	phenoxyethyltrimethylammonium salt condensed with formaldehyde
ASB-3	aminated chloromethylated copolymer of styrene and butadiene

1. Introduction

Among various problems of application of crosslinked polyelectrolytes (CP) in biology, the interaction between CP and ions of organic biologically active substances (BAS) plays one of the central roles. The greatest success in this field has been reached in works where CP specially synthetized for selective and high capacity bonding of BAS is used. In this connection, it has become necessary to study the features of formation and properties of crosslinked polyelectrolytes including the inhomogeneity and mobility of network elements. The equilibrium interaction between CP and BAS is the second important problem in the development of concepts about the role of CP in biology. At the same time, it is important to consider the kinetics of heterogenous mass exchange including the diffusion of BAS ions into polyelectrolyte networks. Among the possible fields of practical application of CP in biology, the immobilization of enzymes and proteins, as well as preparation of highly purified BAS by using CP are considered. Analytical chromatography is not examined here because it is not virtually related to the CP structure. The authors considered it important to pay sufficient attention to the presentation of their own works, while devoting due attention to the investigations of their colleagues working in this field and to alternative views.

2 Structure of Crosslinked Polyelectrolytes (Biosorbents)

2.1 Formation of Crosslinked Structures

The porosity and permeability of CP are the most important factors determining their ability to sorb and immobilize BAS. For solving these problems, it was necessary to synthesize various types of porous and permeable CP differing in the mobility of elements of the crosslinked structure and in the rigidity of the polymer backbone. For biological problems related to the application of CP as biosorbents, it has been found necessary to use CP with a marked structural inhomogeneity.

The formation mechanism of structure of the crosslinked copolymer in the presence of solvents described on the basis of the Flory-Huggins theory of polymer solutions has been considered by Dušek [1, 2]. In accordance with the proposed thermodynamic model [3], the main factors affecting phase separation in the course of heterophase crosslinking polymerization are the thermodynamic quality of the solvent determined by Huggins constant χ for the polymer-solvent system and the quantity of the crosslinking agent introduced (polyvinyl comonomers). The theory makes it possible to determine the critical degree of copolymerization at which phase separation takes place. The study of this phenomenon is complex also because the comonomers act as diluents.

The mechanism of phase separation proposed here (and also observed experimentally) involves the formation in the first stage of polymer "blanks", the globules size depends on the initial comonomers and the copolymerization conditions. In the case of slow phase separation proceeding near the thermodynamic equilibrium of the polymer and solvent, the crosslinked structure is formed from relatively similar supermolecular structures. However, heterophase fluctuations can be considerably increased when the system becomes more distant from the equilibrium state [4].

Inhomogeneous crosslinked structures are formed under all kinetic conditions of reacting comonomers [5]. They acquire quasi-gel structures, macro- and heteroreticular for biosorbents considered below. Inhomogeneities are revealed most distinctly for macroporous copolymers. Depending on the affinity of the solvent for the crosslinked copolymer formed, structures with different degrees of porosity are generated [6, 8]. In the case of a thermodynamically "good" solvent, as a result of low surface tension small microglobules (of the order of magnitude of one nanometer) are formed. In the next stage, a network structure is formed with the distance between the microglobules of the same order of magnitude, which leads to the formation of micropores.

In the case of a thermodynamically "poor" solvent which does not solvate the copolymer, microglobules agglomerate into macroglobules and this leads to the formation of a macroporous structure. In this way, a macroporous St-DVB copolymer has been synthesized [9, 10] and a CP has been obtained by sulfonation [11]. The inert solvent is usually called a pore former. Macroporous polyelectrolytes retain porosity even upon complete dehydration if the amount of the pore former and the crosslinking agent exceeds a certain critical value [7]. A special feature of structure formation in crosslinked polymers is intramolecular crosslinking or cyclization. Cyclization increases the degree of inhomogeneity of the crosslinked polymer at low conversion [12].

For flexible chain copolymers based on acrylic and methacrylic acids (AA and MA) crosslinked with a polyvinyl component, the inhomogeneity of the structures formed depends on the nature of the crosslinking agent, its content in the reaction mixture and the thermodynamic quality of the solvent [13, 14].

In this case, the elements of the crosslinked structure exhibit higher mobility, the permeability of the crosslinked structure depends on the degree of hydration. It should be noted that the pore size in hydrated crosslinked copolymers is determined by small-angle X-ray scattering or with the aid of electron microscopy using special methods of preparation for the CP samples [15].

Table 1. Huggins parameters for a soluble MA-EDMA copolymer (before the gel point) and a linear PMA in acetic acid solutions at 299 K

Solvent% [a] (Acetic Acid)	χ MA-EDMA	χ PMA
5%	1.6	1.2
45%	0.5	0.4
60%	0.8	0.7

[a] Solvent added to monomers.

Table 2. Formation of pre-structures in MA-EDMA (4 mol% of EDMA) mixtures

Solvent%[a] (Acetic Acid)	Conversion (%)	Initial rate	Quantity of C = C Bonds, %		
			At the gel point	in sol fraction after the gel point	in gel at 100% conversion
5	2	0.5	98–100	8–10	8–10
45	15	0.05	65	55	45

[a] Solvent added to monomers.

New structural variants of crosslinked permeable polyelectrolytes, copolymers of AA or MA, and bi-or trivinyl comonomers with vinyl groups relatively distant from each other were obtained by precipitative copolymerization [16–21].

For MA-EDMA copolymers, the thermodynamic quality of the solvent characterized by Huggins parameter χ for the soluble system up to the gel point changes in the same direction as PMA (Table 1) [21].

In free-radical copolymerization in a thermodynamically "poor" solvent which consists of a 5% acetic acid solution (Ac A) for the MA-EDMA system, double bonds of the crosslinking agent are consumed effectively even before the gel point. Copolymerization proceeds at a high rate (Table 2). In a good solvent solvating the copolymer (45% Ac A), copolymerization proceeds slowly and is accompanied by considerable intramolecular crosslinking involving pendant double bonds. However, these bonds are not effectively used even at high conversions. Three-dimensional networks of carboxylic polyelectrolytes formed under these conditions exhibit different structural stability which may be characterized by a change in the degree of CP swelling with the pH shift of the solvent. The data shown in Fig. 1 indicate that crosslinked structures formed in a poor solvent (5 and 60% Ac A) change their volume relatively slowly when the degree of ionization of carboxyl groups (α) increases (in spite of the effect of electrostatic repulsion between the chains). The formation of crosslinked structures in a good solvent solvating the copolymer (45% Ac A) leads to a more traditional picture: a considerable dependence of the relative degree of swelling (K_w) on the degree of ionization of CP.

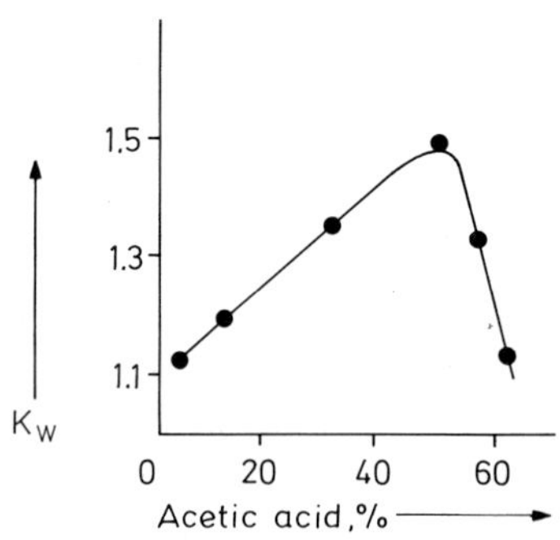

Fig. 1. Steady structure of crosslinked MA-EDMA copolymers obtained in solutions of acetic acid of different concentrations. K_w – relative swelling coefficient, K_{sw} – swelling coefficient (by $\alpha = 0$ and $\alpha = 0.5$, respectively)

$$K_w = \frac{K_{sw}(\alpha = 0.5)}{K_{sw}(\alpha = 0)}$$

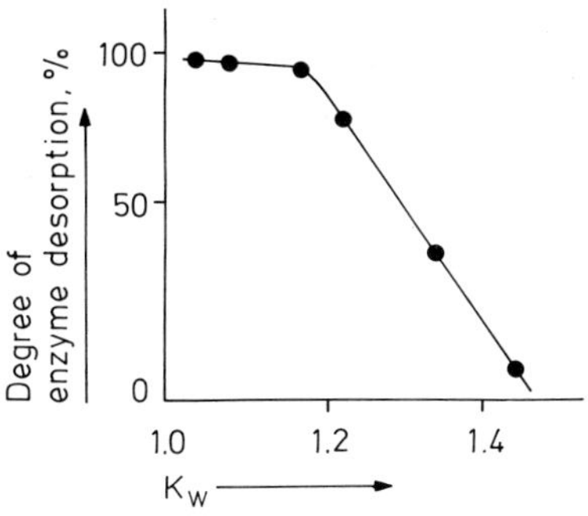

Fig. 2. Reversibility of bonding of terrilytin enzyme to an MA-HHTT heteroreticular copolymer with different steady structure

The steady structure determined by the value of K_w (Fig. 1) for the entire class of carboxylic CP obtained by precipitation copolymerization is one of the most important factors determining the possibility of reversible bonding of proteins absorbed by carboxylic CP with a high sorption capacity [16, 19]. Thus, for the MA-HHTT system (Fig. 2), a complete desorption of enzyme is carried out on crosslinked copolymers characterized by low K_w values. In crosslinked structures exhibiting looser structure ($K_w \gg 1$), owing to the mobility of chain fragments of CP especially in the process of desorption, the macromolecules of sorbed protein are irreversibly captured as a result of a marked polyfunctional interaction.

2.2 Macroreticular and Heteroreticular Polyelectrolytes

The inhomogeneity degree of the CP structure (biosorbents) is an important factor determining a number of their sorption properties with respect to the molecules of biopolymers and other complex organic ions. From this standpoint, it is convenient to follow, for example, the variation in the properties of the copolymers of MA and EDMA obtained by free-radical copolymerization in thermodynamically good and poor solvents with the amount of the divinyl comonomer. Fig. 3 shows that with increasing of the amount of the crosslinking agent, even in a good solvent (30% Ac A solution), the value of the relative swelling coefficient decreases when CP undergoes ionization. In other words, high content of a divinyl monomer favors the formation of relatively rigid crosslinked polymers or with a steady structure. When a thermodynamically poor solvent (5% Ac A solution) is used, (i.e., when precipitative polymerization is carried out) the formation of a CP with a steady structure is much more effective. In these systems (Fig. 3, curves 5 and 6), relative swelling coefficient is low over the entire range of the degrees of ionization. The breaks on curves 5 and 6 show that these structures exhibit additional inhomogeneity. Hence, two groups of CP structures should be singled out. One group with a small content of the crosslinking agent exhibits a high degree of swelling and as will be shown below, has a high permeability for the molecules of biopolymers. The second group may be characterized as having a

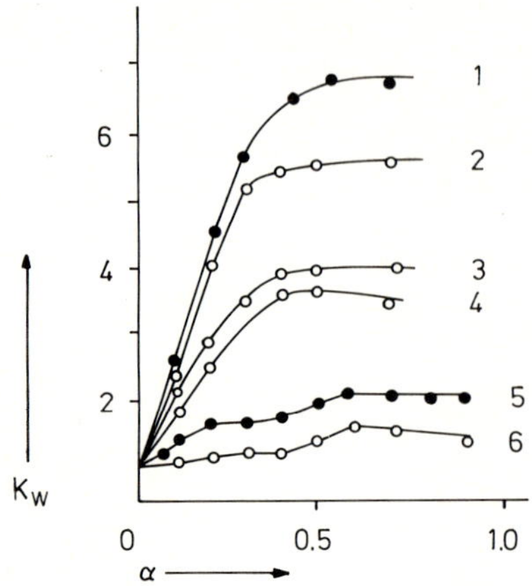

Fig. 3. Change in relative swelling coefficient K_w during ionization of macroreticular MA-EDMA copolymers obtained in 30% Ac A solution (*1–4*) and heteroreticular MA-EDMA copolymers obtained in 5% Ac A solution (*5, 6*) with different amounts of the crosslinking agent – EDMA (mol%): *1*) 1.0; *2*) 1.5; *3*) 2.5; *4*) and *5*) 4.0; *6*) 10.0

more heterogeneous structure and a lower value of the relative degree of swelling. In accordance with this, the first and the second groups of CP may be called macroreticular and heteroreticular polyelectrolytes, respectively.

In addition to the relative degree of swelling, another characteristic may be suggested for evaluating the transition from macro- to heteroreticular polyelectrolytes. According to Flory's method [22], for weakly crosslinked polyelectrolytes it is possible to calculate the average molecular weight of a chain between chemical junctions (M_c) using the relation

$$M_{c\,(theor)} = \frac{2M_m}{f}\left(\frac{1}{N} - 1\right) \tag{2.1}$$

where M_m is the molecular weight of the monomer unit, N is the molar fraction of the crosslinking agent, and f is the functionality of the crosslinking agent.

Table 3. Theoretical and experimental values of M_c for crosslinked MA-EDMA copolymers

EDMA content (mol%)	M_c (exp)		M_c (theor)
	Solvent (acetic acid) used in synthesis		
	5%	45%	
1	–	–	4200
2	2200	2000	2100
4	2200	1100	1000
8	2700	2100	500

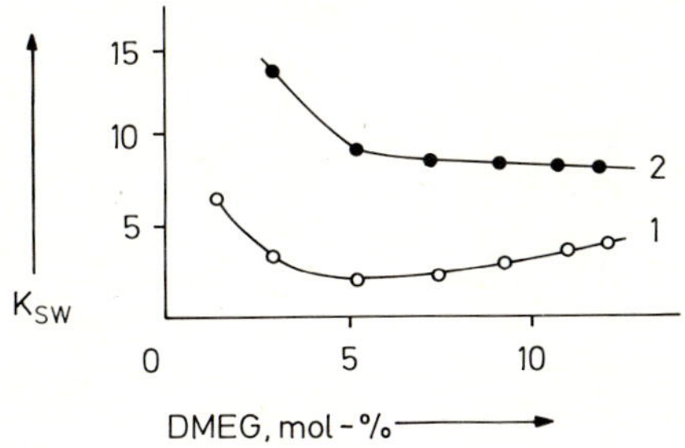

Fig. 4. Swelling coefficients of crosslinked MA-DMEG copolymers as a function of the content of the crosslinking agent: *1)* pH 2.8; *2)* pH 12.5

For the systems considered above, the value of M_c was calculated according to Flory and determined experimentally from the equilibrium degree of swelling using an isopiestic method [23]. The data given in Table 3 show that the theoretical and experimental values of M_c become greatly different when the amount of the crosslinking agent in CP increases. This disagreement is particularly pronounced in systems obtained in a poor solvent (5% Ac A solution). Hence, the above assumption that the formation of heteroreticular copolymers is favored by both poor solvent in precipitation polymerization and the amount of the crosslinking agent introduced is confirmed.

The properties of macro- and heteroreticular polyelectrolytes have been investigated in great detail [15–21, 23–26]. Considerable attention has been devoted to the MA-DMEG system. In this system, like in the system containing EDMA, the swelling coefficients become closer to each other at the degrees of ionization of CP $\alpha = 0$ and 1 when the amount of the crosslinking agent increases (>6 mol%) (Fig. 4). The differences between macro- and heteroreticular structures can be followed from the dependence of the differential enthalpy of ionization on the degree of ionization (Fig. 5). For a macroreticular copolymer obtained with a low content of the crosslinking agent, the value of differential enthalpy gradually

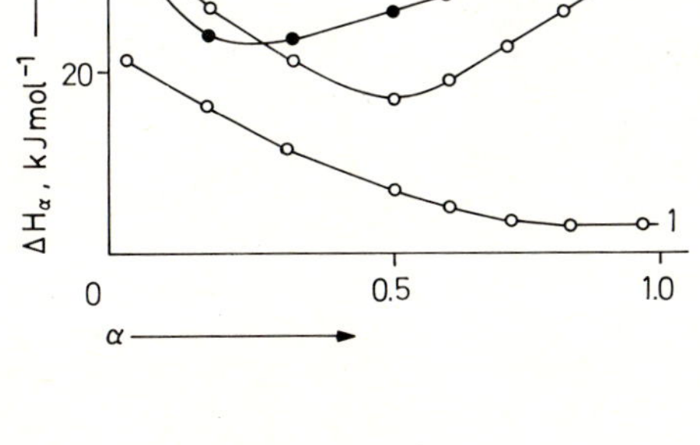

Fig. 5. Differential enthalpy of ionization of crosslinked MA-DMEG copolymers as a function of the degree of ionization of carboxylic groups. Crosslinking agent (DMEG) content (mol%): *1)* 3; *2)* 7.5; *3)* 12

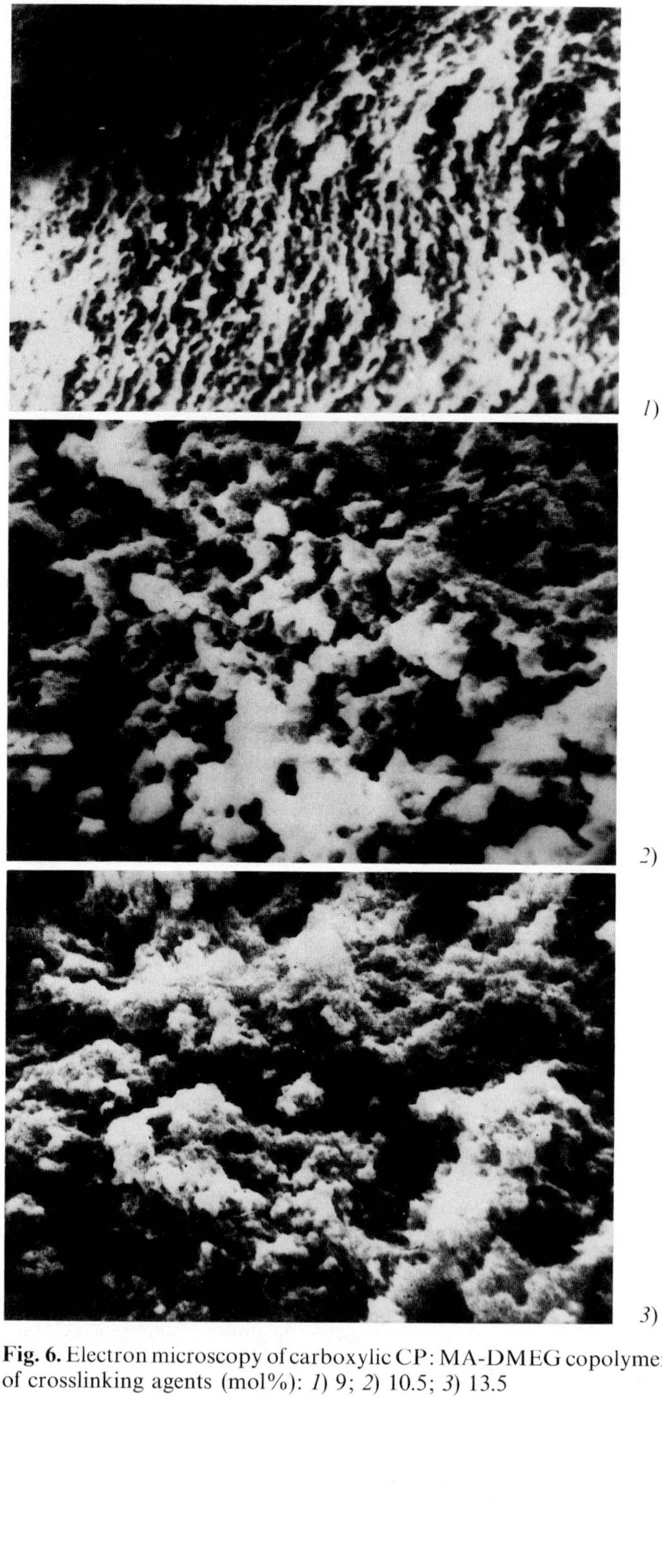

Fig. 6. Electron microscopy of carboxylic CP: MA-DMEG copolymers with different contents of crosslinking agents (mol%): *1*) 9; *2*) 10.5; *3*) 13.5

decreases with increasing α, whereas for heteroreticular copolymers a bimodal shape of curves $\Delta H_\alpha - \alpha$ is observed. This fact confirms that a pronounced inhomogeneity of the crosslinked structure exists in heteroreticular copolymers.

When in precipitative copolymerization aromatic comonomers (derivatives of salicylic or benzoic acid) and HHTT are used as crosslinking agents the heterogeneity of the crosslinked structure greatly increases. This heterogeneity is manifested by increasing swelling with increasing content of the crosslinking agent [27].

Macro- and heteroreticular carboxylic polyelectrolytes lose their porous structure upon high temperature dehydration, this loss being to a certain extent irreversible. For carrying out electron microscopy to establish the inhomogeneity of these groups of CP in the hydrated state, it is convenient to use either consecutive replacement of water first by diethyl ether and then by polymerizable substances capable of filling the voids, or lyophylic drying of the copolymers [28, 29]. The latter method provides the best possibility of evaluating the porosity of initial structures of hydrophilic CP. The microphotographs shown in Fig. 6 illustrate the fact that in the range of 9–13.5 mol% of DMEG, the pore size increases. This increase should affect the permeability during the sorption of biopolymer molecules. In fact, the sorption capacity of MA-DMEG copolymers demonstrates an unusual dependence on the content of the crosslinking agent (Fig. 7). For macroreticular copolymers at a low DMEG content, a well-known trend of decreasing sorption capacity with increasing amount of the crosslinking agent is observed. In contrast, for heteroreticular sorbents the sorption capacity for proteins increases with the amount of the crosslinking agent due to increasing heterogeneity of structure and increasing pore size.

The sorptions of the macromolecules of proteins and some other large organic ions to macro- and heteroreticular polyelectrolytes are comparable. Thus, for an antibiotic of the anthracycline series, rubomycin, the dependence of sorption capacity on the content of the crosslinking agent in CP is also characterized by a curve with a minimum. Moreover, just as in the case of protein sorption, the high capacity sorption of rubomycin on a macroreticular polyelectrolyte (with a small amount of crosslinking agent introduced) is to a considerable extent

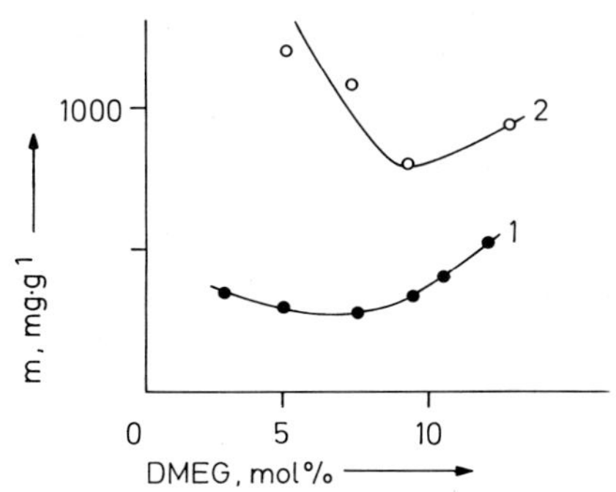

Fig. 7. Quantity of enzymes (m) *1*) terrilytin and *2*) pancreatic ribonuclease, bonded by crosslinked MA-DMEG copolymers with different contents of the crosslinking agent

irreversible in constrast to the high capacity reversible sorption on a heteroreticular copolymer.

It should be noted that the manifestation of macro- and heteroreticulation is also observed for other carboxylic CP. A similar picture was obtained for weakly alkaline CP based on the derivatives of methacrylamide with HHTT [26].

2.3 Mobility of Elements of Crosslinked Structures

The insolubility of crosslinked polymers greatly limits the possibility of their physico-chemical investigation. Until recently, no information existed on the structural organization of chain fragments between chemical and physical junctions and especially on the mobility of these chain fragments. The problems related to chain mobility in CP are of interest mainly for carboxylic CP with a low crosslink density. In these CP, various conformational states of chain parts can be present in the most distinct form. With increasing crosslinking, the possibility of conformational rearrangements should decrease. From this viewpoint, the transition range from macro- to heteroreticular carboxylic polyelectrolytes is of certain interest.

In order to study the mobility of elements of crosslinked structure of CP, it is suitable to use their microdisperse forms [30–35]. On the one hand, in potentiometric titration the equilibrium is quickly attained for these forms and on the other hand the effect of light scattering in spectral methods of investigation (e.g., polarized luminescence) can be greatly decreased.

Sedimentation systems with the particle size of 0.1–1.0 μm were obtained by mechanical dispersion of the copolymers with subsequent washing, fractionation, and separation of fractions. Micrograin forms (gr.) were synthesized by suspension copolymerization and fractionated. For the description of properties of weakly swelling and weakly dissociating gels, Katchalsky [36] has proposed an equation which contains the electrostatic potential $e\psi$

$$\mathrm{pH} = \mathrm{pK}_0 - \log \frac{1-\alpha}{\alpha} + \frac{0.434\, e\psi}{kT} \tag{2.2}$$

The empirical relationship describing the ionization of carboxylic CP of the form of Henderson-Hasselbaech equation has been proposed by Gregor [37]

$$\mathrm{pH} = \mathrm{pK}_\alpha - n \log \frac{1-\alpha}{\alpha} \tag{2.3}$$

The parameter n reflects the measure of deviation of the system from the behavior of the monomeric acid where $n = 1$, i.e., it characterizes the degree of interaction between the neighboring functional groups of the macroion. The value of n depends on the structure of the polyelectrolyte and the nature of the counterion $\mathrm{pK}_\alpha = \mathrm{pK}_0 - \log(1-\alpha)/\alpha$ is the negative decadic logarithm of the effective dissociation constant of the carboxylic CP depending on α.

The characteristic property of linear polyelectrolytes is the existence of a relationship between chain conformation and the electric state of the macromolecule [38]. The physical basis of conformational transitions in macromolecules is the fact that a structuralized state (with a large number of contacts between monomer units) is energetically more advantageous, whereas the state of the unstructuralized (coiled) macromolecules is advantageous with respect to the entropy. The conformational transition is determined by the equality of the Gibbs free energies of the two states. For a family of polyacids of the acrylic series, the change in the free energy in the range of conformational transition is equal to zero for PAA, 0.63 kJ/g-equiv. for PMA and 4.2 kJ/g-equiv. for poly(ethyl acrylic acid) [39].

For weak polyelectrolytes, the conformational state and mobility of chains are due to the contributions of electrostatic and non-electrostatic components. Hydrogen bonds between carboxylic groups, stabilizing the compact PAA structure are destroyed even at low degrees of ionization ($\alpha \sim 0.05$) [40, 41]. The introduction of the methyl (PMA) or the ethyl group in the polyacid increases the tendency towards structuralization as a result of hydrophobic interactions between these groups. The structure is retained until the energy of electrostatic repulsion of ionized carboxylic groups becomes lower than the energy of hydrophobic interaction stabilizing the ordered structure [42, 43]. The increase in the degree of ionization is accompanied by a transition of the polyacid from the structuralized state into the disordered coiled state. In this case, the size of the macromolecule and the chain mobility increase cooperatively. The cooperative conformational transition in the poly(ethyl acrylic acid) occurs at higher degrees of ionization than in PMA [39].

For microdisperse forms of weakly crosslinked copolymers based on MA, regardless of the nature of the crosslinking agent, titration curves exhibit a conformational transition induced by pH changes which is analogous to the trans-conformational transition during ionization of linear PMA [30, 31] (Fig. 8). In the range of conformational transition (in a $pK_\alpha - \alpha$ system of coordinates), a constancy of pK_α values of PMA in the degree of ionization range of 0.1–0.4 has been observed. This fact indicates that at low degrees of ionization local

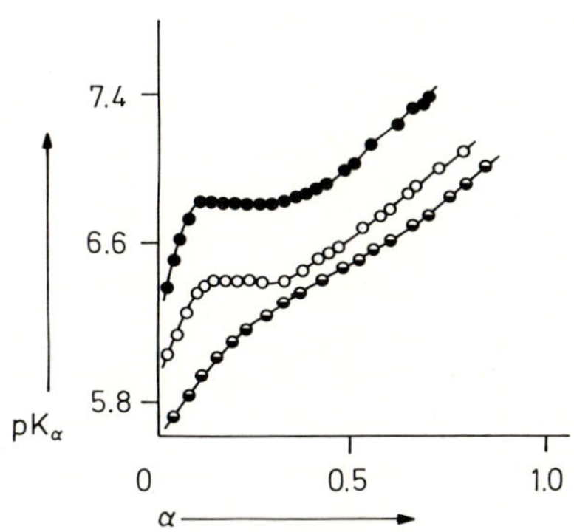

Fig. 8. Titration curves of *1)* macroreticular AA-EDMA (2.5 mol%) copolymer; *2)* linear PMA and *3)* macroreticular MA-EDMA (2.5 mol%) copolymer

Table 4. Values of $pK_{\alpha=0.5}$, pK_c (of conformational transition) and the range of degrees of ionization in which conformational transition ($\alpha_1 - \alpha_2$) takes place for the carboxylic CP at various ionic strength (C_{NaCl})

Polyelectrolyte	C_{NaCl} (N)	$pK_{\alpha=0.5}$	pK_c	$\alpha_1 - \alpha_2$
PMA	0.001	6.6	6.4	0.15–0.35
MA-EDMA (2.5 mol%)	0.001	7.0	6.8	0.13–0.36
MA-EDMA (2.5 mol%)	0.01	6.7	6.4	0.18–0.29
MA-DVB (2 mol%)	0.001	6.8	6.7	0.12–0.37
MA-DVB (2 mol%)	0.01	6.6	6.4	0.18–0.32

ordered chain structure exists and these structures are destroyed at $\alpha = 0.4$. With increasing ionization of carboxylic groups a monotonic increase in pK_α value in weakly crosslinked AA-EDMA copolymer is observed. The ionic strength of the solution decreases the electrostatic field formed in a carboxylic CP. With increasing ionic strength of the solutions, the start of the conformational transition α_1 is displaced towards higher degrees of ionization (Table 4).

The possibility of conformational changes in chains between chemical junctions for weakly crosslinked CP in ionization is confirmed also by the investigation of the kinetic mobility of elements of the reticular structure by polarized luminescence [32, 33]. Polarized luminescence is used for the study of relaxation properties of structural elements with covalently bonded luminescent labels [44, 45]. For a microdisperse form of a macroreticular MA-EDMA (2.5 mol% EDMA) copolymer (Fig. 9a, curves 1 and 2), as compared to linear PMA, the inner structure of chain parts is more stable and the conformational transition is more distinct. A similar kind of dependence is also observed for a weakly crosslinked AA-EDMA (2.5 mol%) copolymer (Fig. 9b, curves 4 and 5).

Suspension copolymerization in a hydrophobic medium is known to lead to the formation of micrograins partly swelling during ionization as a result of the formation of a denser layer on the grain surface [46, 47]. Since a micrograin copolymer is inhomogeneous, carboxylic groups exhibit different ionization capacity. In heteroreticular micrograin MA-EDMA (2.5 mol%) (gr) (Fig. 9a, curve 3), a larger hindrance to chain motion in ionization is observed, the conformational transition is less cooperative and the maximum chain mobility is attained only at $\alpha > 0.6$. This fact may be related to dense regions of the heteroreticular structure. For linear polyelectrolytes the character of titration curves is determined by the mutual arrangement of charged groups on the polymer chain. In the case of crosslinked analogues, the mutual arrangement of the polymer chains fixed by crosslinks, depend on the conditions of the crosslinked structure formation, which an important and in many cases, a determining effect.

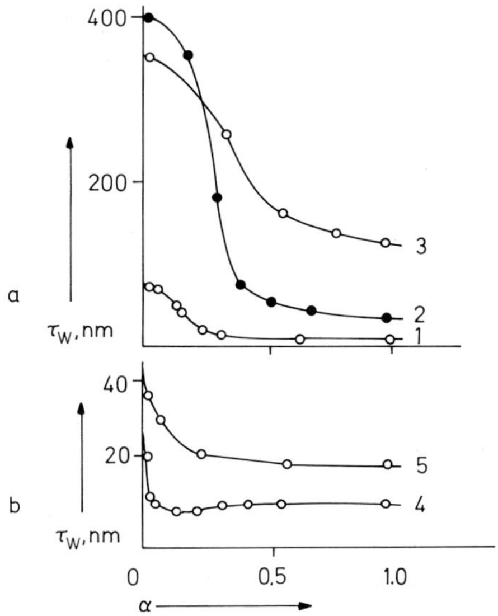

Fig. 9 a, b. Effect of degree of ionization of polyelectrolytes on relaxation time (τ_w) for: **a.** *1*) linear PMA; *2*) crosslinked MA-EDMA (2.5 mol%) copolymer; *3*) MA-EDMA (2.5 mol%) bead copolymer. **b.** *4*) linear PAA; *5*) crosslinked AA-EDMA (2.5 mol%) copolymer

As mentioned above, in heterophase copolymerization the heterogeneity of CP increases with an increase in the amount of the crosslinking agent and decrease in the solvating ability of the solvent. As a result of inhomogeneous distribution of the polymer mass in carboxylic CP, breaks characterized by a constant value of pKα appear on the titration curves of macrodisperse forms at anomalously high degrees of ionization ($\alpha > 0.6$) for the conformational transition of PMA. This range of conformational transition increases with increasing heterogeneity which reflects the disappearance of intramolecular contracts in densely crosslinked regions at high degrees of ionization (Fig. 10). Hence, it is possible to estimate the degree of electrochemical heterogeneity from titration curves of heteroreticular carboxylic polyelectrolytes. Since all carboxylic groups in heteroreticular polyelectrolytes are accessible for titration, it is possible to determine the contribution

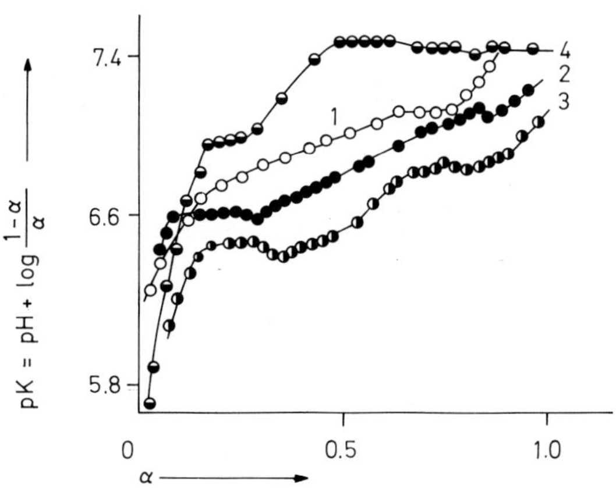

Fig. 10. Titration curves of heteroreticular copolymers: *1*) MA-EDMA (2.5 mol%)gr., synthesized from a 30% Ac A solution; *2*) MA-EDMA (4 mol%), from a 30% Ac A solution; *3*) MA-EDMA (4 mol%), from a 5% Ac A solution; *4*) MA-EDMA (10 mol%), from a 5% Ac A solution

of groups that belong to dense regions of the crosslinked structure. The MA-EDMA (2.5 mol%) copolymer is relatively homogeneous. For a copolymer with 4 mol% of EDMA synthesized in a good solvent (30% Ac A), up to 10% of the titrated groups are located in dense regions. At 5% Ac A and 4 mol% EDMA, 20% of carboxyl groups are located in heterogeneous regions. In the case of a carboxylic CP with 10 mol% of the crosslinking agent, the fraction of the electrochemically inhomogeneous parts attains 50%.

In terms of Henderson-Hasselbalch equation, linear parts with different slopes are observed (Fig. 11). These slopes reflect single stages of titration of microdispersion of carboxylic CP [30, 31]. For a MA-EDMA macroreticular copolymer (Fig. 11-2), the range of $\alpha = 0$–0.15 corresponds to the titration of the PMA exhibiting an inner structure of chain parts, $\alpha = 0.15$–0.4 is the conformational transition and at $\alpha > 0.4$ the ionization of carboxyl groups with disordered coiled state takes place. In the titration of a microdispersion of heteroreticular carboxylic polyelectrolyte, a break in the straight-line dependence occurs at $\alpha \sim 0.6$, the range corresponding to the destruction of the interchain contacts in dense parts (Fig. 11-4).

The results of titration of AA-EDMA copolymers with 2.5 and 10 mol% crosslinking agent are shown for comparision in Fig. 11 (1 and 3). In both cases breaks in the straight-line dependences are observed at low degrees of ionization (0–0.15). They may be caused by weakening of hydrogen bonding as a result of ionization of carboxyl groups. Also, the intramolecular chain mobility obtained by polarized luminescence (Fig. 9b-5) changes. In the case of the AA-EDMA (10 mol%) heteroreticular copolymer, another break in the range of $\alpha \sim 0.6$ is observed. This break may correspond to the start of destruction of interchain contacts in dense regions fixed by crosslinks.

The investigation of structural dynamics of CP is particularly topical in connection with the establishment of correlation between local intramolecular mobility and the reactivity of chain fragments. It has been established that groups located in the most mobile parts of the polymer chain exhibit the greatest reactivity [48]. The chemical heterogeneity in relationship to local mobility is particularly important for interaction with proteins.

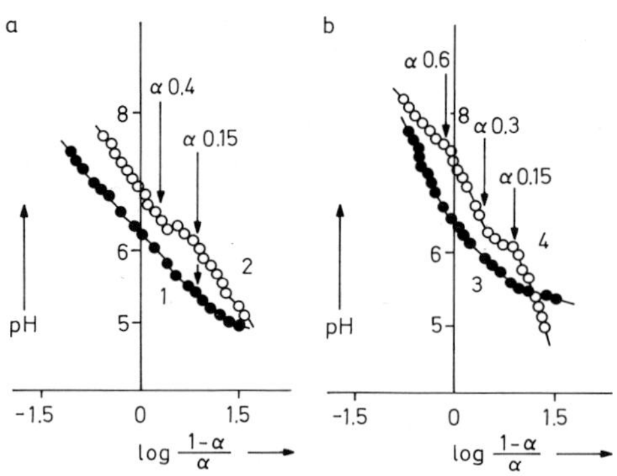

Fig. 11a, b. Titration of microdisperse CP in a linear system of coordinates; **a**) macroreticular copolymers: *1*) AA-EDMA (2.5 mol%); *2*) MA-EDMA (2.5 mol%); **b**) Heteroreticular copolymers: *3*) AA-EDMA (10 mol%); *4*) MA-EDMA (10 mol%)

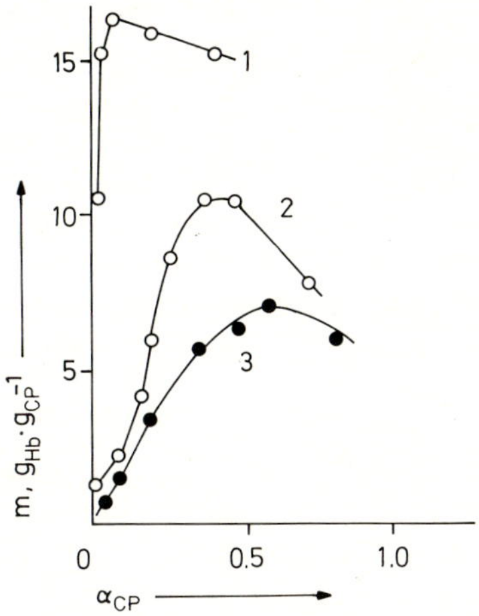

Fig. 12. Effect of the degree of ionization of microdisperse carboxylic CP on binding of haemoglobin: *1*) AA-EDMA (2.5 mol%); *2*) MA-EDMA (2.5 mol%); *3*) MA-EDMA (2.5 mol%)gr. m – quantity of Hb (g) bonded by crosslinked copolymers (g)

For the study of protein sorption on highly permeable carboxylic CP, it is convenient to use as a sample a well soluble protein, haemoglobin (Hb), which is a dipolar ion having the isoelectric point at pH 6.8. Protein bonding to carboxylic CP is determined by the ionization of CP and the electrochemical state of the protein macromolecule. It follows from Fig. 12 that the amount of Hb sorbed by macroreticular AA-EDMA and MA-EDMA copolymers and heteroreticular micrograin MA-EDMA-gr. (in all cases with 2.5 mol% EDMA) increases with the ionization of carboxylic CP and attains the maximum value at the degrees of ionization corresponding to the highest mobility of the elements is the crosslinked structure (at $\alpha = 0.1$, 0.4 and 0.6, respectively). In the case of a macroreticular MA-EDMA copolymer, the curve of the dependence relating the sorption of Hb to the degree of ionization of carboxylic CP (at $\alpha = 0.1-0.4$) reflects the cooperative conformational transition in polymer chains, whereas that for a heteroreticular micrograin copolymer extends over a wide range with a maximum at $\alpha \sim 0.6$. This fact shows that the process of destruction of local structures is not of a very cooperative character in the ionization of –COOH groups in the range of $\alpha = 0.1-0.6$.

As has been shown in Ref. [49], the character of bonding of Hb by other macro- and heteroreticular carboxylic CP reflects the structural and electrochemical features of synthetic copolymers.

3 Sorption Selectivity of Organic Biologically Active Ions by Crosslinked Polyelectrolytes

3.1 Energetic and Entropic Factors of Interphase Ion Exchange

Selectivity of sorption of organic ions by crosslinked polyelectrolytes in competition with small ions, in particular with metal ions, should be considered on the basis of the analysis of thermodynamic relationships of ion exchange.

In the thermodynamic analyses of heterogeneous mass exchange of ions most often binary ion exchange is considered. For using the methods of formal thermodynamics in similar (even the simplest) systems, it is necessery to introduce the concept of ion concentration in the bonded state — in the polyelectrolyte. The above considered experimental results demonstrated that the inhomogeneity of CP — biosorbents require the use of additional ideas on the state of components (solvated exchanging ions). In this case, the energetic inhomogeneity of ionogenic groups is superposed on the morphological inhomogeneity of polyelectrolyte networks [50–52]. Also, in ion exchange the degree of hydration of exchanging ions and CP as a whole varies. Most of these complications may be excluded or greatly decreased if combining counterions and fixed ions on CP and including the nearest environment including hydrating water molecules is considered as a component.

Within the framework of the proposed model, the ion exchange equilibrium between the solution and CP ion exchanger can be described by the equation:

$$dG_{T,P} = \bar{\mu}_1\, dn_1 - \mu_2\, dn_2 + \mu_1\, dn_1 - \bar{\mu}_2\, dn_2 = 0 \tag{3.1}$$

where G is the Gibbs free energy of the system, $\bar{\mu}_1$ and $\bar{\mu}_2$ are the chemical potentials sorbed ions, μ_1 and μ_2 are the chemical potentials of ions in solution and n_1 and n_2 are the numbers of equivalents of ions.

Equation (3.1) does not take into account the electric work because equivalent ion exchange is considered. Therefore

$$\begin{aligned} z_1\, dn_1 &= -z_2\, dn_2\,, \qquad z_1\, d\bar{n}_1 = -z_2\, d\bar{n}_2 \\ dn_1 &= -d\bar{n}_1\,, \qquad dn_2 = -d\bar{n}_2 \end{aligned} \tag{3.2}$$

where z_1 and z_2 are ion charges.

Taking into account the interdependence between chemical potentials and component activity $\bar{\mu}_i = \bar{\mu}_i^0 + RT \ln \bar{a}_i$, Eqs. (3.1) and (3.2) yield a well known ion exchange equation

$$\frac{\bar{a}_1^{1/z_1} a_2^{1/z_2}}{a_1^{1/z_1} \bar{a}_2^{1/z_2}} = K = e^{-\Delta G^0/RT} \tag{3.3}$$

where K is the thermodynamic constant of ion exchange. Therefore, the concentration ratios are given by the equation

$$\frac{m_1^{1/z_1} c_2^{1/z_2}}{c_1^{1/z_1} m_2^{1/z_2}} \frac{\bar{\gamma}_1^{1/z_1} \gamma_2^{1/z_2}}{\gamma_1^{1/z_1} \bar{\gamma}_2^{1/z_2}} = K \tag{3.4}$$

where

$$\frac{m_1^{1/z_1}}{m_2^{1/z_2}} \frac{c_2^{1/z_2}}{c_1^{1/z_1}} = K_s\,.$$

Here γ_1 and γ_2 are the activity coefficients of ions in solution, $\bar{\gamma}_1$ and $\bar{\gamma}_2$ are the coefficients of resin activity, c_1 and c_2 are ion concentrations in solution, m_1 and m_2 are fixed ion concentrations (exchange or weight concentrations) and K_s is the concentration constant of ion exchange, the selectivity constant.

The selectivity of ion exchange K_s can easily be determined experimentally for the simplest systems of exchange between monovalent ions. The value of K_s may be used for analysis of thermodynamic functions ΔG^0, ΔH^0 and ΔS^0, of sorption selectivity

$$\Delta G^0 = -RT \ln K = -RT \ln \int_0^1 K_s \, d\bar{N}_i$$
$$\Delta H^0 = \frac{\partial \ln K}{\partial T}, \qquad \Delta S^0 = \frac{\Delta H^0 - \Delta G^0}{T} \tag{3.5}$$

where $\bar{N}_i$ is the molar fraction of ion in CP.

For more complex systems, e.g., for the exchange of ions with different charges equations similar to Eq. (3.5) are used [53].

In the bonding of complex organic ions by densely crosslinked polyelectrolyte, not all active centers of ion exchanger are accessible to counterions [54]. For these systems Eq. (3.4) becomes

$$\frac{m_1^{1/z_1} c_2^{1/z_2}}{c_1^{1/z_1}(M - m_1)^{1/z_2}} \, \frac{\bar{\gamma}_1^{1/z_1} \gamma_2^{1/z_2}}{\gamma_1^{1/z_1} \bar{\gamma}_2^{1/z_2}} = K \tag{3.6}$$

where M is the maximal exchange capacity of CP for an organic ion (second component). Experimental data show that the value of M is sufficiently constant for many systems under investigation.

The first quantitative investigations of the relationship between CP and organic ions with the simplest structure, quaternary ammonium bases [55], have demonstrated a considerable increase in the sorption selectivity of organic ions (in competing interaction with metal ions) upon passing to ions with a more complex structure. This selectivity increases, for instance, with an increase in the chain length of aliphatic side groups or an passing from an aliphatic to an aromatic substituent in the quaternary ammonium ion.

The thermodynamic analysis of the selectivity of ion exchange with the participation of ions of quaternary ammonium bases [56–58] has shown that an increase in bonding selectivity, when metal ions are replaced by organic ions, which is usually accompanied by an increase in entropy of the system (Table 5). It follows from Table 5 that a drastic increase in bonding selectivity upon passing to a triethylbenzylammonium counterion (the most complex ion) is due to a considerable increase in the entropy of the system.

The sorption selectivity for more complex ions (e.g., ions of antibiotics) in competition with small ions increases to a still greater extent [59]. Particularly high constants for selective bonding (up to some hundreds and thousand) have

Table 5. Thermodynamic analysis of the substitution of ions of quaternary ammonium bases by sodium ions
CP – product of polycondensation of sulfophenol with formaldehyde

Organic counterion	K_s	ΔG kJ mol^{-1}	ΔH kJ mol^{-1}	$T\Delta S$ kJ mol^{-1}
$(CH_3)_4N^+$	6.9	−4.6	−5.9	−1.3
$(CH_3)_3C_2H_4OHN^+$	9.1	−5.4	−11.3	−5.0
$(C_2H_5)_3C_6H_5CH_2N^+$	162.0	−13.4	10.9	24.3

been obtained for novobiocin, an antibiotic of the acid type, on a CP anion exchanger (containing additional aromatic groups) in chloride form [60, 61].

The thermodynamic analysis of these systems played an important role in the interpretation of these data and of the high selectivity. It was found that selective sorption of complex organic ions is accompanied by an increase in the entropy of the system (Table 6).

A comparison of the effects of increasing selectivity of sorption of organic counterions and the entropy control of the replacement of small ions by large organic ions in CP is a rational explanation of these phenomena.

First, apart from ionic interactions non-ionic interactions between complex counterions and CP play a certain role. In a number of systems hydrophobic interactions play an important role in the selectivity of bonding of organic counterions [62].

In some cases, an alternative explanation is possible. It may be assumed that any very complex organic counterion can also interact with the CP matrix with the formation of weak non-ionic bonds, e.g., dipole-dipole bonds or other types of weak interactions. If the energy of these weak additional interactions is on the level of the energy of the thermal motion, a set of microstates appears for counterions and the surrounding CP matrix, which leads to an increase in the entropy of the system. The changes in Gibbs free energy of this interaction may be evaluated in a semiquantitative way [15].

Table 6. Selectivity of ion exchange bonding of antibiotics on crosslinked polyelectrolytes

Antibiotic	second counterion	CP-ion exchanger	K_s	ΔH kJ mol^{-1}	$T\Delta S$ kJ mol^{-1}
Chlortetracycline	Na	SBS	124	7.1	18.4
Oxytetracycline	Na	SBS	110	−15.4	−3.2
Oxytetracycline	H	KRS-4	40	−7.9	1.3
Morphocycline	H	KRS-4	39	0	8.8
Morphocycline	H	KRS-4-T-40	36	6.7	15.5
Penicillin	Cl	ASB-3	19	18.4	25.5
Novobiocin	Cl	FAF	955	16.7	33.5
Novobiocin	Cl	AV-16	725	0	16.3
Novobiocin	Cl	AV-17	1820	0	18.4

3.2 Interaction between Crosslinked Polyelectrolytes and Dipolar Ions

Dipolar ions simultaneously bearing positively and negatively charged ionogenic groups are aminoacids, peptides, proteins, nucleotides, nucleic acids and many other organic ions like antibiotics of the tetracycline group. Taking amino acids as an example, a peculiar mechanism of interaction between dipolar ions and CP in aqueous solutions near the isoelectric point has been demonstrated [59]. Sulfostyrene CP in the H-form bind amino acid from aqueous solutions of glycine or alanine without desorption of hydrogen counterions into solution. If the Na-form of this CP is used, the absorption of aminoacids decreases by one order of magnitude and more. The explanation of this phenomenon lies in the fact that the dipolar ion of the amino acid $H_3N^+RCOO^-$ is bound mainly in the form of a cation by the following mechanism

$$R_1SO_3^-H^+ + H_3N^+RCOO^- \rightarrow R_1SO_3^-N^+H_3RCOOH\,. \qquad (3.7)$$

This behavior has been observed to a certain extent for some proteins at a low ionic strength of the solution. In contrast to this mechanism, in buffer systems

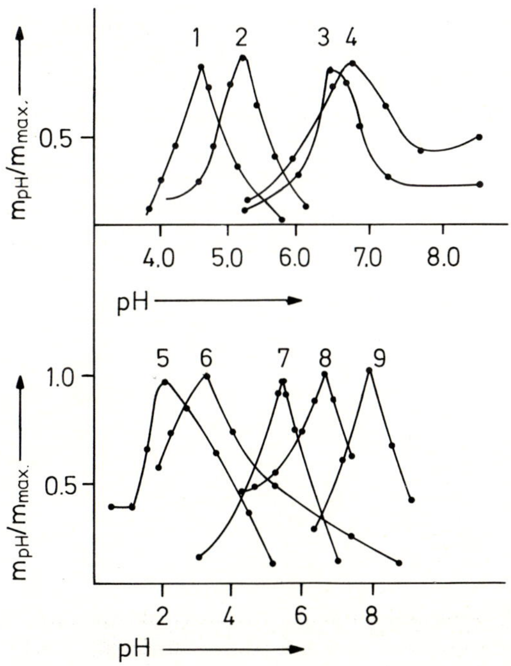

Fig. 13. Relative sorption capacity of proteins by carboxylic CP Biocarb-T vs pH of solution: *1*) terrilytin, *2*) insulin, *3*) chymotrypsinogen, *4*) pancreatic ribonuclease, *5*) pepsin, *6*) thymarine, *7*) thermolysine, *8*) haemoglobin, *9*) lysozyme. m_{max} – quantity of protein bonden on Biocarb-T by pH_{max}

Table 7. Thermodynamic functions of the interaction between proteins and heteroreticular highly permeable Biocarb-T-biosorbent (MA-HHTT copolymer)

Protein	ΔG (kJ mol^{-1})	ΔH (kJ mol^{-1})	$T\,\Delta S$ (kJ mol^{-1})
Ribonuclease	−3.3	−2.0	1.3
Serum albumin	−4.0	5.4	9.4
Terrilytin	−5.5	0	5.5
Insulin	−5.3	−4.4	0.9
Pepsin	−4.5	−4.9	9.4

with a relatively high ionic strength, a typical ion exchange sorption of various dipolar ions like proteins is observed.

High sorption capacities with respect to protein macromolecules are observed when highly permeable macro- and heteroreticular polyelectrolytes (biosorbents) are used. In buffer solutions a typical picture of interaction between ions with opposite charges fixed on CP and counterions in solution is observed. As shown in Fig. 13, in the acid range proteins are not bonded by carboxylic CP because the ionization of their ionogenic groups is suppressed. The amount of bound protein decreases at high pH values of the solution because dipolar ions proteins are transformed into polyanions and electrostatic repulsion is operative. The sorption maximum is either near the isoelectric point of the protein or depends on the ratio of the pI of the protein to the $pK_{\alpha=0.5}$ of the carboxylic polyelectrolyte [63]. It should be noted that this picture may be profoundly affected by the mechanism of interaction between CP and dipolar ions similar to that describedby Eq. (3.7).

Since the mechanism of interaction between proteins polyfunctional with respect to ionogenic groups and CP is complex, an approximate method of calculation of sorption selectivity constants according to the inverse form of Langmuir isoterm should be used. Hence, the approximate values of ΔG, ΔH and ΔS obtained from Eq. (3.5) should be applied (Table 7).

Calculations were usually carried out under the conditions of a pH-maximum of protein bonding. The experimental results show that the interaction of proteins and most other complex organic substances with CP is accompanied by an increase in the entropy of the system.

3.3 Cooperative Effect

In accordance with Eq. (3.4) or Eq. (3.6), the concentration selectivity of ion exchange is variable depending on the degree of ideality of the solution and CP phase. For dilute solutions at a constant ionic strength, it is possible to take into account as a variable only the degree of non-ideality of the CP phase. For the systems considered here, it is convenient to study the effect of the molar fraction of organic counterions ($\bar{N}_i$) on the concentration selectivity constant. Fig. 14 shows the dependences of K_s on the molar fraction of oxytetracycline in CP. For CP

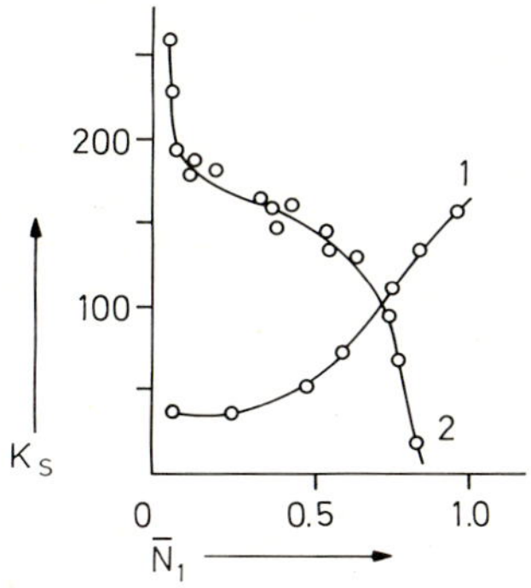

Fig. 14. Selective bonding of oxytetracycline vs its molar fraction ($\bar{N}_i$) in Dowex 50 × 1 sulfonated cation exchangers with different exchange capacity: *1*) 5.0 mg-equiv/g; *2*) 2.3 mg-equiv/g

with a low number of ionogenic groups at a low degree of filling by an organic counterion (oxytetracycline), high K_s values are observed. This fact may be explained by ion-ionogenic interactions with the CP matrix, namely, with units bearing no ionogenic groups. A further drop in the K_s value is a consequence of the consecutive substitution of metal ions with ions of the organic substance on functional groups with decreasing selectivity which corresponds to the above concents of their energetic inhomogeneity. Quite a different picture is observed for CP with a large number of ionogenic groups. The initial low selectivity can be explained by the absence of centers of additional interaction with organic counterions located near the ionogenic group. A very important role is played by a fact which seems unusual in the simplest consideration but is quite frequent for organic counterions: an increase in the concentration selectivity constant, K_s, with increasing degree of filling of CP with an organic ion. This phenomenon is due to a cooperative effects in intermolecular interactions between neighboring organic counterions on CP. The mechanism of interaction of organic counterions with each other in the CP phase may involve different types of bonds. In particular, on a Biocarb-T biosorbent the cooperative bonding of oxytetracycline, oleandomycin and novocain is due to hydrophobic interactions. As shown in Fig. 15, the transition from an aqueous to methanol solution cancels the cooperativity effect almost completely. At present, the phenomenon of cooperative interaction between CP and organic counterions has been studied in sufficient detail [24, 62, 64, 65]. Of particular interest is the study of the cooperative interaction between CP, especially carboxylic CP, and protein macromolecules. In the interaction with protein, the presence of methyl groups and the high concentration of weakly dissociating carboxylic groups of these CP favor the formation of hydrogen bonds and hydrophobic contacts. The specificity of carboxylic CP is the possibility of controlling the ratio of these interactions by changing the degree of dissociation of carboxylic groups [63]. The bonding of proteins to macro- and heteroreticular carboxylic CP is often characterized by a sigmoidal shaped graph, which shows a dependence on the relative amount of bonded protein (m/M) at its equilibrium concentration in solution (C_{eq}) (Fig. 16). S-shaped curves usually refer to ligand-induced conformational changes in systems in which the interaction between a

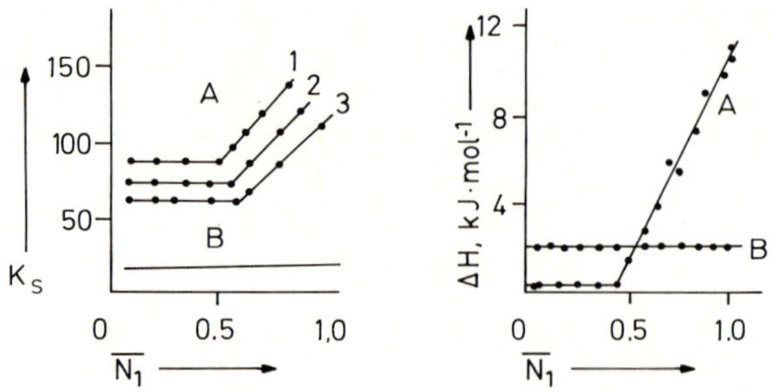

Fig. 15. Cooperative effect of bonding of organic ions on Biocarb-T biosorbent: *A*. 0.1 N NaCl in water; *B*. 0.1 N NaCl in methanol. For *1*) novocain, *2*) oleandomycin and *3*) oxytetracycline

molecule and a ligand with one active center on the carrier macromolecule modifies the neighboring center increasing its affinity for the ligand and thus, favoring the attachment of subsequent molecules. The cooperative process characterizes the interaction between active centers [66].

For the quantitative description of the cooperative process in the macromolecule-low molecular weight ligand systems, Hill's equation is used. It expresses the dependence of the degree of macromolecule saturation with the ligand (Y) on the equilibrium concentration of the ligand in solution [67]:

$$\frac{\mathrm{Y}}{1-\mathrm{Y}} = \mathrm{KC}_{\mathrm{eq}}^{\mathrm{n}} \tag{3.8}$$

where the exponent n characterizes the measure of cooperativity of the process and its value corresponds to the hypothetical quantity of ligand molecules required

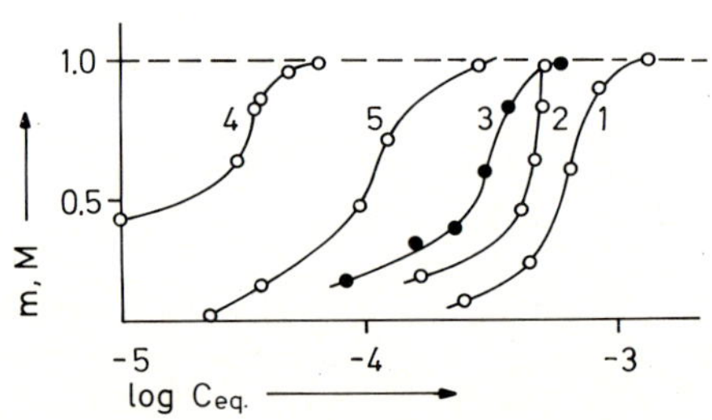

Fig. 16. Isotherms of protein bonding by nonionized CP ($\alpha = 0$): *1*) lysozyme – MA-EDMA (2.5 mol%) copolymer; *2*) haemoglobin – MA-EDMA (2.5 mol%) copolymer; *3*) haemoglobin – AA-EDMA (2.5 mol%) copolymer; *4*) haemoglobin – MA-EDMA (2.5 mol%)gr. copolymer; *5*) serum albumin – MA-EDMA (2.5 mol%) copolymer. C_{eq} (mol/l) – concentration of proteins in solution by equilibrium with CP; $\frac{\mathrm{m}}{\mathrm{M}}$ – relative sorption capacity of proteins on CP (capacity by definite concentration of proteins to limited capacity)

for the bonding to proceed by the cooperative mechanism. The isotherm is S-shaped ($n > 1$) and reflects the cooperative process in which the bonding of the next molecule is preferable in the presence of one already sorbed. At $n = 1$, the bonding centers are assumed to be identical and independent, and Eq. (3.8) describes the Langmuir type isotherm. The parameter n called Hill's constant is determined by the slope of the cooperative curve in coordinates $\log Y/(1 - Y)$ vs. $\log C_{eq}$ in the range of CP saturation degrees $Y = 0.2$–0.8. The maximum cooperative interaction is found near Y 0.5 for a symmetrical S-shaped isotherm. The initial and final parts of the curve correspond to the bonding of the first and last ligand molecules which are characterized by the value of n close to unity.

By analogy, Hill's equation may be used for a quantitative study of the cooperative interaction between a protein (ligand) and a carboxylic CP (carrier) with the formation of protein clusters in the CP phase [68]. In the interaction between proteins and carboxylic CP, the role of the single bonding center is played by a system of functional groups of both components. Hence, the formation of protein-weakly crosslinked carboxylic CP complex may be represented by analogy with linear polyelectrolytes [69, 70] as the sticking of parts of carboxylic CP chains to the surface of the protein globule.

The degree of saturation of carboxylic CP with protein (Y) is determined by the ratio of the amount of protein bonded under these conditions (at a predetermined concentration in solution) to the maximum amount $Y = m/M$. In this case, Hill's equation becomes

$$\frac{m}{M} = \frac{KC_{eq}^{n}}{1 + KC_{eq}^{n}} . \tag{3.9}$$

Figure 17 shows in Hill's coordinates the bonding isotherm of proteins: haemoglobin (Hb), serum albumin (SA) and lysozyme (Ls) by a crosslinked MA-EDMA (2.5 mol%) copolymer in the non-ionized ($\alpha = 0$) and partly ionized forms ($\alpha = 0.13$ and $\alpha = 0.4$) reflecting different levels of structural organization of chains in carboxylic CP. In all cases, the cooperative character of protein bonding to macro- and heteroreticular granular samples (obtained by suspension polymerization) is revealed. In Hill's system of linear coordinates, two parts can be distinguished with different slopes (n_I and n_{II}) corresponding to the two stages of the process: the non-cooperative stage when the filling of CP chains with protein is low ($n_I \sim 1$), which at a certain degree of filling (m/M) passes into the cooperative stage ($n_{II} \gg 1$) (Table 8). The value of n may be treated as the number of bonding centers for protein macromolecules acting as cooperative units and locally fixed on CP in space. In this connection, the value determined by the number of monomer units of MA per molecule of bonded protein (P) can characterize in the first approximation the density of filling of CP chains with protein (M_{MA}/M_P) [69]. If it is assumed that the sorbed protein molecules are located along the chain close to each other, then the contour length of the part occupied by one protein

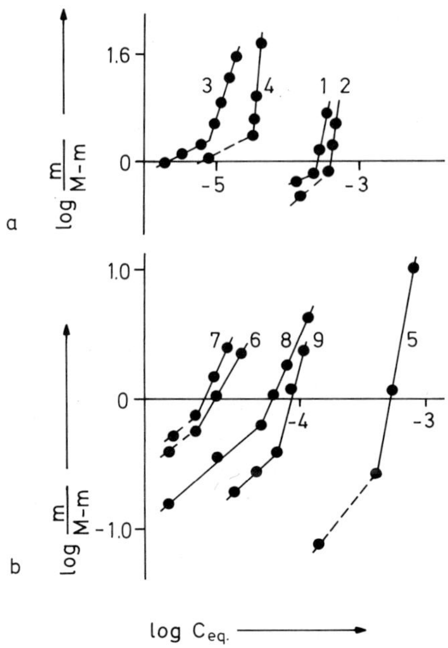

Fig. 17a, b. Isotherms of protein bonding (in Hill's coordinates) by crosslinked polyelectrolytes. **a**) Haemoglobin bonding: *1*) AA-EDMA (2.5 mol%), $\alpha = 0$; *2*) MA-EDMA (2.5 mol%), $\alpha = 0$; *3*) MA-EDMA (2.5 mol%), $\alpha = 0.4$; *4*) MA-EDMA (2.5 mol%)gr., $\alpha = 0$. **b**) Lysozyme bonding: *5*) MA-EDMA (2.5 mol%), $\alpha = 0$; *6*) MA-EDMA (2.5 mol%), $\alpha = 0.13$; *7*) MA-EDMA (2.5 mol%), $\alpha = 0.4$. Serum albumin bonding: *8*) MA-EDMA (2.5 mol%), $\alpha = 0.13$; *9*) MA-EDMA (2.5 mol%), $\alpha = 0.4$

molecule is comparable with the diameter of the protein globule. If a flexible polymer chain winds round the protein molecule, the protein can occupy the contour length of the chain corresponding to one turn of the polymer chain around the protein globule.

It should be remembered that the density of filling of chains of carboxylic CP with lysozyme at $\alpha = 0$ (when local structural chain order exists) is determined by the value $M_{MA}/M_P = 56$. This corresponds to the interaction along the entire chain and is probably due to the disordering of the inner structures of CP chains by the negatively charged Ls — macroion. At $\alpha = 0.13$ and 0.4, the most dense packing of protein molecules in carboxylic CP is achieved ($M_{MA}/M_P = 24$ and 16, respectively). It has been found by polarized luminescence that the distribution of Ls in the matrix of carboxylic CP is irregular and associates are formed [72, 73]. The achievement of a high selectivity of interaction between proteins and carboxylic CP may be explained by a cooperative association of protein globules.

The assumption of the association of Hb in the pores of carboxylic cation exchangers has been advanced in Ref. [47] on the basis of electron microscopy: at the maximum filling, almost all the pore surface is filled with Hb associates which are ordered star-shaped structures. Interprotein interaction in the adsorption immobilization of enzymes have been reported in Refs. [74, 75].

Table 8. Parameters of cooperative bonding of lysozyme, haemoglobin and serum albumin (proteins) by a macroreticular MA-EDMA (2.5 mol%) copolymer

Protein	Calcul. M_{MA}/M_P [a] turn	diam.	α_{CP}	Stage I n_I	Stage I M_{MA}/M_P	I–II Y	Stage II n_{II}	Stage II M_{MA}/M_P
Ls (pI = 10.5)			0.00	0.7	250	0.22	4.0	*56*
R_{st} = 2.06 nm	51	16	0.13	0.8	*57*	—	—	*24*
			0.40	1.5	*38*	0.44	2.5	16
Hb (pI = 6.8)			0.0	1.3	880	0.41	8.0	360
R_{st} = 3.1–3.4 nm	80–85	25–27	0.13	0.8	200	0.62	3.2	120
			0.40	0.7	*66*	0.52	3.0	*34*
SA (pI = 4.7)			0.0	—	—	—	2.3	160
R_{st} = 3.6 nm	90	29	0.13	1.0	320	0.33	3.1	*85*
			0.40	1.2	560	0.15	3.5	*106*

[a] The length of monomer unit of MA is taken to be 0.25 nm [71]. The length of the polymer chain equal to the turn or the diameter of the protein globule is calculated taking into account the hydrodynamic size of the protein molecule via Stockes radius (R_{st}).

By using Hill's coefficient, it is possible to draw a conclusion about the character of the process and to determine ligand concentration in one cooperative unit.

The method of determination of integral values of changes in Gibbs free energy in cooperative processes has been proposed and theoretically based by Wyman in the general theory of linked functions [76] and applied to the study of cooperative homo- and heterotropic interactions between biomacromolecules with different low molecular weight ligands. According to Wyman's theory, the probability for a center on the carrier molecule to be bonded to a ligand is considered to be a function of ligand activity required for a predetermined degree of saturation of the carrier. Hence, apart from Hill's constant, an important characteristic of the cooperative process is the equilibrium concentration of the ligand determined under the conditions of maximum manifestation of the cooperative effect ($Y \sim 0.5$ for a symmetric cooperative curve). This ligand activity (C_m) determines the measure of changes in the stability of the cooperative complex, i.e., the changes in Gibbs free energy

$$\Delta G = RT \ln C_m . \tag{3.10}$$

Table 9 gives the values of ΔG and ΔG_n calculated according to Wyman's theory.

The theory of linked functions establishes the general thermodynamic meaning of the cooperative behavior of the system. On the basis of Hill's equation, $n \equiv \dfrac{d \ln Y/(1 - Y)}{d \ln C_{eq}}$, Wyman proposed an expression for the change in Gibbs thermodynamic potential characterizing the interaction between the bonding

Tabelle 9. Values of ΔG and ΔG_n for cooperative bonding of proteins by an MA-EDMA (2.5 mol%) macroreticular copolymer in the variation of the degree of ionization of carboxylic groups

Protein	α_{CP}	Log C_m	ΔG (kJ mol^{-1})	n_{II}	$(\zeta_2 - \zeta_1)$	ΔG_n (kJ mol^{-1})
Lysozyme	0.00	−3.26	−18.2	4.0	0.86	3.0
	0.13	−4.54	−26.0	—	—	—
	0.40	−4.56	−26.8	2.5	0.24	0.6
Haemoglobin	0.00	−3.37	−19.1	8.0	0.42	1.5
	0.13	−4.30	−24.4	3.0	0.42	1.5
	0.40	−4.80	−27.8	1.3	0.14	0.5
Serum albumin	0.00	−4.00	−22.7	2.3	0.62	2.2
	0.13	−3.90	−22.0	3.1	0.34	1.2
	0.40	−3.90	−22.0	3.5	0.34	1.2

centers in the cooperative process (ΔG_n)

$$\Delta G_n = RT \int_{Y=0}^{1} \left(1 - \frac{1}{n}\right) \frac{dY}{Y(1 - Y)} = RT \int_{C_{eq}}^{\infty} (n - 1)\, d \ln C_{eq} \tag{3.11}$$

or

$$\Delta G_n = RT \sqrt{2}\, (\zeta_2 - \zeta_1), \tag{3.12}$$

where $(\zeta_2 - \zeta_1)$ is the difference between the coordinates in the new geometric plot [76].

Table 9 shows that the value of ΔG_n of the cooperative interaction between bonding centers is within the error in the determination of integral ΔG values. This fact can either indicate the slight mutual influence of the centers or be caused by the compensation between the enthalpy and entropy components of Gibbs free energy.

The change in the degree of filling of carboxylic CP with protein is reflected in the character of contributions of ΔH and $T\Delta S$. The method of differential microcalorimetry was used for the investigation of the thermal effects at varying α in the Hb-CP (Table 10).

In the interaction between Hb and the non-ionized form of macroreticular carboxylic polyelectrolytes ($\alpha = 0$), the enthalpy (which is considerable in the absolute value) sharply changes its sign from negative to positive at high degrees of filling $Y = 0.8$–1.0 [77]. The existence of the internal structure in the chains of CP investigated here at $\alpha = 0$ suggests that the energetically favorable process of protein bonding at low degrees of filling (at the most active centers) proceeds without deformation of CP chains. Close to the limiting saturation, the change

Table 10. Change in differential enthalpy in the interaction between Hb and CP at different degrees of CP filling with protein (Y)

CP	α_{CP}	Y	ΔH (kJ mol^{-1})
AA-EDMA (2.5 mol%)	0.0	0.11	−17.1
	0.0	0.24	−26.0
	0.0	0.80	14.6
	0.0	1.00	27.3
AA-EDMA (2.5 mol%)	0.2	0.05	−19.1
	0.2	0.60	−32.0
	0.2	1.00	−41.1
MA-EDMA (2.5 mol%)	0.0	0.10	−14.2
	0.0	0.22	−16.4
	0.0	0.83	57.1
	0.0	1.00	62.3
	0.3	1.00	−28.1
	0.45	1.00	−62.0
MA-EDMA (2.5 mol%) – gr. (suspension copolymerization)	0.00	0.14	0
	0.00	0.60	−22.1
	0.00	1.00	−38.0

in sign of the energetic component may be related to a disordering of the chain structure under the influence of a strong local electrostatic field of the sorbed protein molecule.

With increasing CP ionization, increasing chain mobility and the disordering of the local structure, the saturation of CP with protein in all cases becomes

Table 11. Thermodynamic functions of interaction between Hb and CP as a function of the degree of CP ionization

CP	α_{CP}	ΔG (kJ mol^{-1})	ΔH (kJ mol^{-1})	T ΔS (kJ mol^{-1})	ΔS (J mol^{-1} K^{-1})
AA-EDMA (2.5 mol%)	0.00	−19.7	27.3	47	158
	0.20	−29.0	−41.1	−12.1	−41
MA-EDMA (2.5 mol%)	0.00	−19.0	62.3	81.1	273
	0.30	−25.4	−28.1	−2.7	−9
	0.45	−29.0	−62.0	−33.0	−111
MA-EDMA (2.5 mol%) gr. (suspension copolymerization)	0.0	−26.6	−38.2	−11.6	−39

energetically favorable. The changes in ΔH are similar for heteroreticular polyelectrolyte (obtained by suspension polymerization) for which a set of energetically unequal carboxylic groups exists as a result of the inhomogeneity of the cross-linked structure (Chapter 2.3).

The change in ΔH is compensated for by the change in entropy of the system (Table 11).

The determining contribution of energy to the process of Hb bonding is very pronounced only in those cases when CP chains exhibit an internal structure. Hence, it may be assumed that the entropy factor is due to the disordering of the local chain structure of carboxylic CP as a result of polyfunctional interaction with protein, which is favored by the increase in the quantity of the micro-states of the system. In partially ionized CP, rearrangements of flexible chain parts on the protein globule are possible and the process is governed by the energetic component.

The relationship of thermodynamic functions of selective bonding of Hb to a series of carboxylic CP in the variation of the degree of ionization of carboxylic groups is expressed by the effect of enthalpy-entropy compensation (Fig. 18). The compensation effect of enthalpy and entropy components is the most wide-spread characteristic of many reactions in aqueous solutions for systems with a cooperative change in structure [78].

As a result of thermodynamic analysis it is shown that protein bonding to carboxylic CP exhibiting a local internal chain structure is determined by the entropy factor, whereas, if the arrangement of flexible chain parts on the protein globule is possible, the energetic component predominates.

Cooperative effects are of considerable interest for high capacity chromatography of BAS, since for practical purposes high-selectivity bonding is possible only in cooperative processes. This is very important for carrying out the sorption, separation and concentration of BAS.

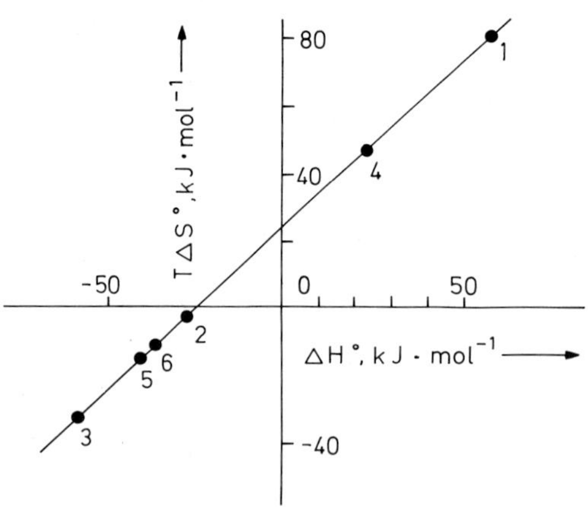

Fig. 18. Compensation dependence of thermodynamic functions ΔH^0 and $T\,\Delta S^0$ of haemoglobin bonding by carboxylic CP: *1)* MA-EDMA (2.5 mol%), $\alpha = 0$; *2)* MA-EDMA (2.5 mol%), $\alpha = 0.13$; *3)* MA-EDMA (2.5 mol%), $\alpha = 0.4$; *4)* MA-EDMA (2.5 mol%), $\alpha = 0$; 5) MA-EDMA (2.5 mol%), $\alpha = 0.2$; *6)* MA-EDMA (2.5 mol%) gr., $\alpha = 0$

3.4 Microdisperse Forms of Crosslinked Polyelectrolytes

A quasi-homogeneous model for CP with a counterion was considered in Sections 3.1 and 3.2 as a model suitable for thermodynamic analysis. However, the energetic inhomogeneity of functional groups was taken into account. A system of homogeneous parts of CP was considered in the form of a large set of microparts. The validity of this concept is confirmed by sufficient reproducibility of thermodynamic functions when CP grains of different dimensions are used [60]. These concepts, however, are not valid any longer if the CP grains are reduced to micron size. The experimental data have shown that on passing from the grains generally used for ion exchange sorption of organic ions or their immobilization having a diameter of ~200 μm to microdisperse grains (d ~ 10 μm) the selectivity coeffi-

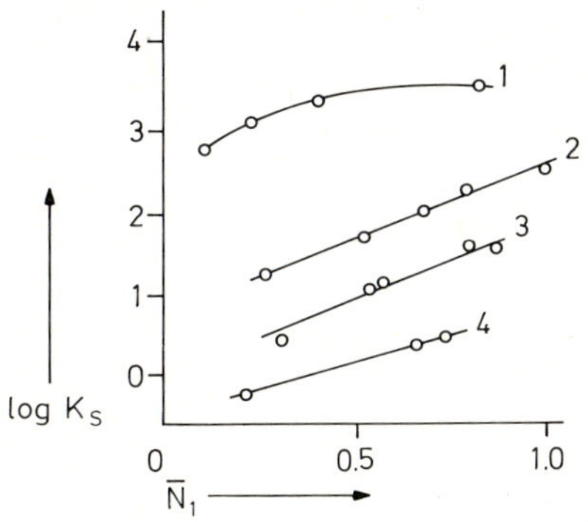

Fig. 19. Selective bonding of novobiocin by crosslinked polyelectrolytes ARA-3p-T-40 (*1* and *4*) and ARA-1p-T-40 (*2* and *3*) with different grain sizes *1*) and *2*) 10 μm *3*) and *4*) 200 μm. $\bar{N}_i$-molar fraction of novobiocine in the ion exchanger

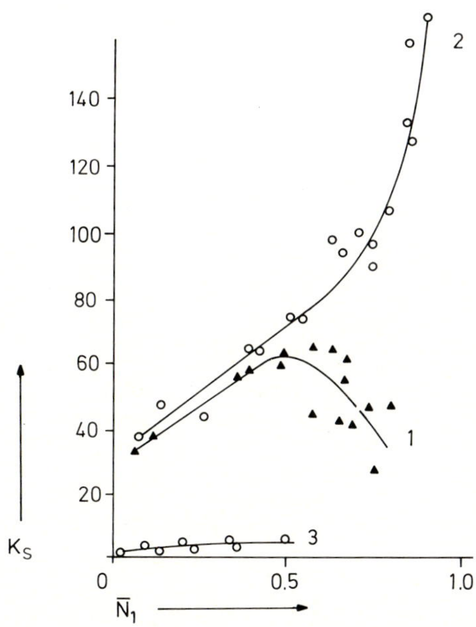

Fig. 20. Selective bonding of oxytetracycline by KRS-5p cation exchanger with different grain sizes: *1*) 90 μm, *2*) 7 μm, *3*) 0.1 μm

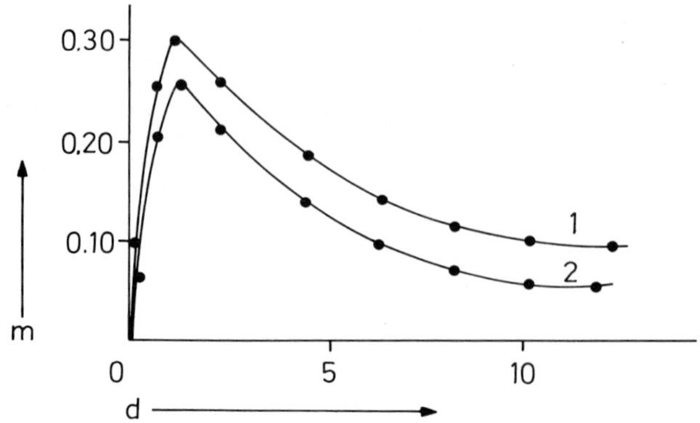

Fig. 21. Influence of grain size (d) of CP *1*) FAV-17 and *2*) FAV-16 on precipitation (m, g/g) of serum albumin. $C_{SA} = 10^{-4}$ mol/l, $C_{CP} = 0.05$ mg/g, pH 6.2

cients, e.g., with respect to novobiocin antibiotic, increase by one or three orders of magnitude [79] (Fig. 19). The weaker but still important effect of oxytetracycline [80] is observed in the dispersion of resin sulphonate from 90–7 µm (Fig. 20). However, further dispersion of grains to 0.1 µm leads to a considerable decrease in the bonding selectivity for organic ions. This effect should be compared with the following well-known fact: the existence of lower selectivity for organic ions on linear soluble polyelectrolytes rather than on similar crosslinked CP [64].

In addition it should be added that microdisperse forms of CP can precipitate proteins from solutions. Figs. 21–23 show that CP microdispersions with particle size of 1–2 µm precipitate serum albumin from solutions [81] in complete agreement with general flocculation laws for polyelectrolytes. The figs. show an extreme

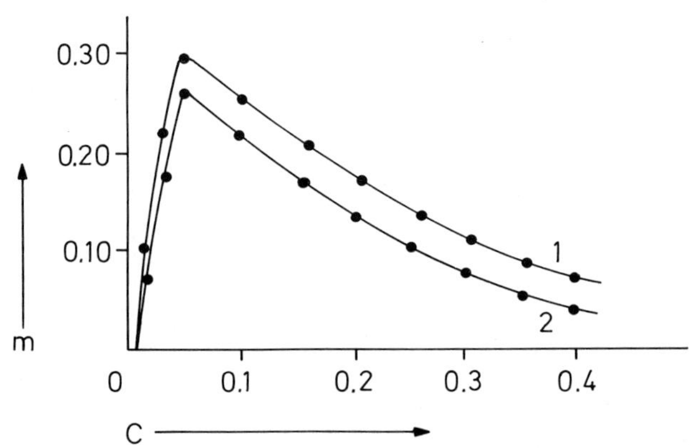

Fig. 22. Influence of concentration of *1*) FAV-17 and *2*) FAV-16 microdispersions on precipitation of serum albumin. $C_{SA} = 10^{-1}$ mol/l, pH 6.2

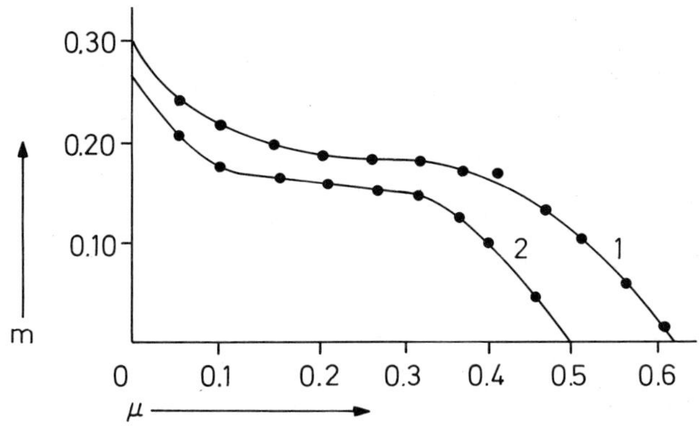

Fig. 23. Influence of ionic strength of solution (μ) on precipitation of serum albumin by *1)* FAV-17 and *2)* FAV-16 microdispersions. $C_{SA} = 10^{-4}$ mol/l, $C_{CP} = 0.09$ mg/ml, pH = 6.2

dependences for SA precipitation upon the change in concentration and the size of CP (precipitant) grains. The figs. also show the curves (with a plateau region and a break) of the dependence of the degree of protein precipitation on the ionic strength of the solution typical of the flocculation phenomena.

The interpretation of the relationships obtained here is based on the same principles of polyfunctional interaction between CP and organic ions which are considered in sections 3.1–3.3. The dispersion of CP grains to a certain size (1–10 μm) yields particles retaining the ability of polyfunctional interaction with organic ions. Simultaneously with increasing dispersion, the mobility of elements of the crosslinked structure also increases, which favors additional interaction. Further dispersion of CP (d ~ 0.1 μm) gives so weak networks that the spatial effect of polyfunctional interaction with organic ions drastically decreases similar to linear polyelectrolytes [64].

It should be noted that the flocculation effect produced by microdisperse forms of CP is applicable not only to protein solutions but also to solutions containing various organic ions.

4. Protein (and Enzymes) Immobilization by Crosslinked Polyelectrolytes

With the development of the polymer field in medicine, great attention has been paid to particulate forms of drugs [82]. The most widespread methods for the preparation of particulate drugs are microencapsulation and microgranulation, i.e., the inclusion of BAS into spherical shapes of predetermined dimensions. One form of particulate drugs is microcapsules or artificial cells as they were called by the founder Chang [83, 84]. However, the inclusion of a BAS (enzyme) solution into a polymer capsule from a semipermeable membrane leads to considerable diffusion limitations for the substrate [85]. Another variety of microencapsulation is the inclusion of a drug into lyposomes phospholipid vesicles capable of being consumed in the organism [86]. Polyaldehyde microspheres are used for the covalent attachment of BAS to the surface [87].

All the existing methods of immobilization involve formation of a high local BAS concentration and retention of its biological activity. In this respect, the use of disperse forms of CP as carriers of BAS used for different purposes is very promising [88]. In this case, the CP-protein interaction is an important factor in controlling the structure and properties of these systems.

The high permeability of carboxyl CP in combination with their considerable selectivity of bonding protein macromolecules have made it possible to use them as carriers for the immobilization of proteins and enzymes with the aim of protection against some physiological factors (e.g., the pH of the medium).

4.1 Enzyme Immobilization

The immobilization of enzymes with the formation of insoluble forms is usually intended for the development of specific catalysts for technical purposes. Here, we consider another medico-biological problem of the preparation of insoluble enzymatic systems based on crosslinked polyelectrolytes, used in the replacement therapy for oral administration.

In the framework of this problem, the questions of the stability of enzymes in acid solutions imitating the acid medium of the stomach have mainly been studied. An attempt to obtain enzyme-polymer complexes dissociating with the isolation of the active enzyme upon an increase in the pH of the surrounding solution has also been made [88–94]. As mentioned above, heteroreticular carboxylic CP with high content at crosslinking agent-biosorbents of the Biocarb type, exhibit exceedingly high sorption capacity with respect to enzyme ma-

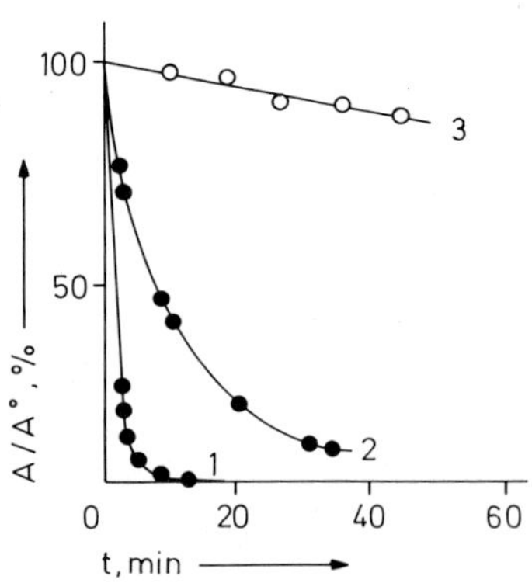

Fig. 24. Kinetics of acid inactivation of α-amylase (*Bac. subtilis*) in solution (*1*, *2*) and immobilized on Biocarb (*3*) *1*) pH 2; *2*) and *3*) pH 4. A/A_0 is the value of relative enzymatic activity (compared to the initial activity A_0 before inactivation), substrate — starch

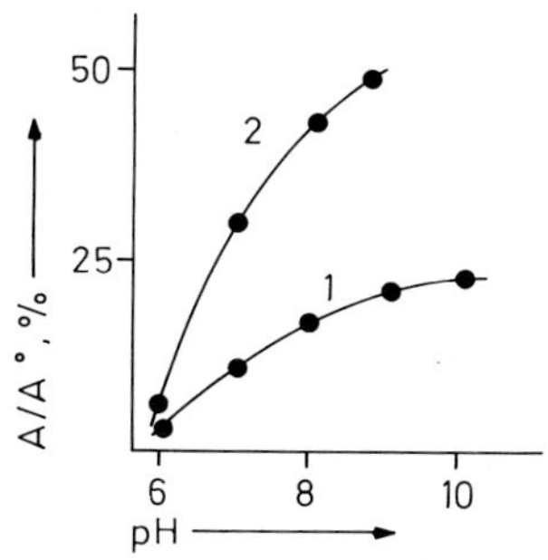

Fig. 25. Reactivation of inactivated of *1)* α-amylase at different pH values of the solution, *2)* α-amylase immobilized in Biocarb

cromolecules. The polyfunctional interaction between protein molecules and these biosorbents determines the stabilization effect of native protein conformation. Fig. 24 shows inactivation curves for α-amylase in solution and in the immobilized state.

The reactivation of enzymes (after their partial inactivation in an acid medium) upon passing into a medium of pH 8 is also of great importance for oral use (Fig. 25). Enzymes immobilized in crosslinked polyelectrolytes are characterized by a structural memory even after considerable inactivation. Under changed conditions, this leads to a considerable or almost complete reactivation of the enzyme, whereas in the reactivation of a free enzyme in solution under similar conditions the enzymatic activity is restored on a lower level.

The stability of the enzyme-polymer complex and its dissociation upon the variation of pH depends on the structural and other physico-chemical properties of CP and enzyme molecule. Thus, a Biocarb-T heteroreticular biosorbent (Fig. 26) is characterized by a stability of its complex with α-amylase (under the condition of its stabilization) in acid solutions and a complete dissociation of the complex during isolation of the active enzyme at pH 7–8.

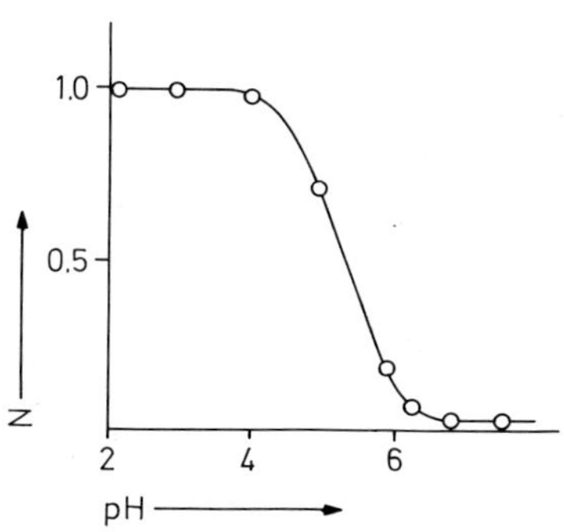

Fig. 26. Dissociation of the α-amylase – Biocarb complex. N is the fraction of dissociated (active) α-amylase

4.2 Haemoglobin Immobilization

Highly selective bonding of a protein from red cells – haemoglobin (fulfilling the most important gas-transporting function of blood) by microdisperse forms of carboxylic CP has made it possible to attain high local concentrations of protein molecules without any distortion of the native structure by sorption immobilization. In the microparticles of the Hb–CP complex, the weight amount of bonded Hb can exceed that of the polymer carrier by a factor of 10–15. Mild bonding conditions have made it possible to retain or increase the functional activity in haemoglobin immobilized in the microdisperse form [95].

The functional activity of Hb is related to the efficiency of oxygen transport $Hb + 4\,O_2 \rightleftarrows HbO_8$ carried out in the organism in a narrow physiological range of the partial pressure of oxygen in blood pO_2 between lungs and tissues (shaded area in Fig. 27). It is expressed quantitatively by the dependence of the degree of oxygenation of Hb on the pO_2 of the medium and is of a cooperative character determined by the parameter n. This is physiologically important because it allows the organism to react rapidly to the smallest change in pO_2 in tissues [96, 97]. When Hb is obtained by a haemolysis of red cells, its allosteric effectors are lost, and the gas transport of Hb in solution decreases in comparison with that of Hb in the red cell. It can be seen from Fig. 27 that the curve is displaced to the left: the value of pO_2 decreases upon half-saturation P_{50}. When Hb is liberated from the red cell the super organization of haemoglobin peculiar to Hb molecules contained in the red cells is distorted [98].

Figure 27 and Table 12 show that the characteristics of oxygen transport of the microdisperse form of the immobilized haemoglobin (Im Hb) are virtually identical with the functional parameters of red cells: the cooperativity of interaction between subunits in the Hb moleculed upon immobilization (n) and the ability of allosteric regulation (Bohr-effect) are completely retained. The most important fact is the increase in the efficiency of oxygen release in the case of Im Hb up to the red cells level as compared to haemoglobin P_{50}. It may be assumed that the association of Hb molecules observed in the cooperative interaction with carboxylic CP is capable of affecting the biological activity of Hb, and the polymer matrix is a kind of polymer allosteric effector improving its gas transport properties.

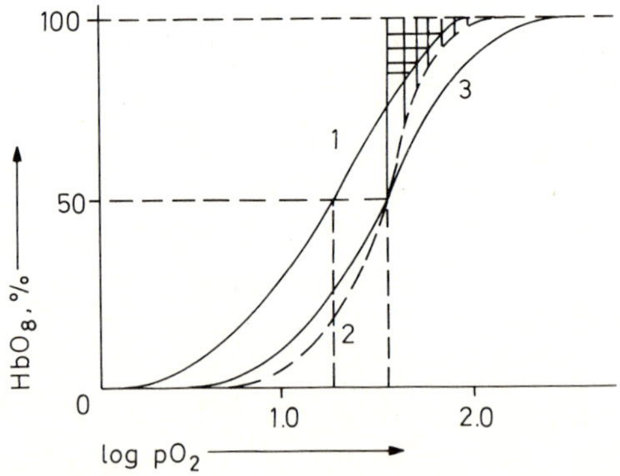

Fig. 27. Dissociation curves for: *1*) oxyhaemoglobin in solution; *2*) oxyhaemoglobin in human red cells and *3*) immobilized haemoglobin. 37 °C, pH 7.4; pCO_2 40 Torr, pO_2 0–150 Torr

Table 12. Parameters of reversible oxygen transport for haemoglobin and a microdisperse form of immobilized haemoglobin (pH 7.4, 37 °C, pCO_2 = 40 Torr, pO_2 = 0–150 Torr)

Sample	P_{50}	n	$\frac{d \log P_{50}}{d\ pH}$
Human haemoglobin	15–18	2.6	−0.45
Human red cells	26–28	2.7	—
Im Hb	25–28	2.5	−0.42

The influence of pH on the affinity of Hb for oxygen known as the Bohr-effect indicates that protons retain the allosteric regulation of oxygen transport. It is also an indirect confirmation of the ability of Hb and Im Hb for transporting carbon dioxide. The values of the Bohr-effect d log P_{50}/d pH for Hb and Im Hb are close to each other in the pH range 7.1–7.4. It is possible that the effect of the micro-environment of carboxylic CP on immobilized Hb and its polyfunctional interaction represents the interaction between Hb and the structural elements inside the red cell [99].

The high gas transport efficiency of Im Hb, the high local Hb concentration in a particle of microne dimensions make it possible to consider Im Hb as a system representing distinct characteristics of the red cell.

The aggregate stability of Im Hb fraction with the particle size of 5–10 μm and the suspensions of human red cells with the size of 5–8 μm washed from the plasma were investigated simultaneously in solutions of different ionic strengths (from 0.1–0.6 N NaCl). Similar curves were obtained for red cells and Im Hb. The presence of a maximum on the curves of the aggregate stability is due to the possible recharging of the double electric layer as a result of a strong screening effect of the electrolyte. The decrease in the stability of both systems with increasing salt concentration higher than 0.25 N is probably caused by the compression of the double electric layer characterizing the distribution of counterions around the charged particle. A similar behaviour of these systems in solutions upon variation in ionic strength may explain the identity of the charge sign of Im Hb and red cells. For the confirmation of this assumption, the electrophoretic mobility of Im Hb particles ~1 μm in size in various media is compared with that of human red cells in Table 13. The value of electrophoretic mobility of red cells (U) coincides with the data for many types of intact cells [100]. The particles of Im Hb (just as red cells) bear a negative charge the value of which virtually coincides with that of red cells (σ).

Preliminary biological tests showed the compatibility of Im Hb with blood and the theoretical possibility of intravenous injection and functioning in the organism. The use of microparticles of Im Hb with a covalently bonded marker permitted the determination of the time of microparticle circulation in the blood channel of rats. After 7 h. of observation, up to 30% of the introduced amount of Im Hb was retained in the blood of the animals.

Table 13. Electrophoretic mobility (U) and surface charge density (σ) of particles, Im Hb in various media

Sample	Medium	U $\left(\frac{\mu m\ S^{-1}}{V\ cm^{-1}}\right)$	σ (ESU cm^{-2})
Human red cells	0.15 N NaCl	−1.1	3300
Im Hb	0.15 N NaCl	(−1.2)–(−1.5)	3000–3600
Im Hb	Blood Plasma in 0.15 N NaCl (1 : 10)	−1.0	2800
Im Hb	Blood plasma in 0.15 N NaCl (1 : 5)	−1.1	3100
Im Hb	0.75% of albumin in 0.15 N NaCl	−1.1	3300
Im Hb	4.5% of albumin 0.15 N NaCl	−0.8	2500
Im Hb	0.75% of gelatinol in 0.15 N NaCl	−0.7	–
Im Hb	4.5% of gelatinol in 0.15 N NaCl	−0.5	–

5. Mass-Exchange Kinetics of Organic Ions on Crosslinked Polyelectrolytes

5.1 Permeability of Crosslinked Polyelectrolytes for Organic Ions

The rate of mass-exchange for any counterions between the solution phase and the crosslinked polyelectrolyte (ion exchanger) may be determined by one of the transport stages: by convective transport in the solution phase (at the interface between the ion exchanger and the solution), the quasi-diffusion transfer in the CP phase and the chemical act of ion exchange on the functional groups of exchangers. The ion exchange process with the participation of organic counterions is usually limited by quasi-diffusion in the crosslinked polyelectrolyte [54, 101]. This quasi-diffusion is a complex process of ion diffusion through the most permeable weakly crosslinked parts or channels which are not always located in the radial direction. This process is subsequently combined with diffusion in highly crosslinked parts of the networks. One of the more outstanding examples of the combined diffusion process in the compositional ion exchanger is considered in Section 5.2. The picture of the mass-exchange is very complex, the quasi-diffusion of ions in a crosslinked polyelectrolyte is virtually always represented by radial diffusion and characterizes the averaged kinetic permeability of the crosslinked material. In some cases, the effect of the electric field on quasi-diffusion is also considered [102, 103].

The diffusion of ions into a spherical particle may be described, for low values of the Fourier number $\left(\frac{\bar{D}t}{R^2} < 0.15\right)$, by the equation

$$F = \frac{Q_t}{Q_\infty} \approx \frac{6}{R}\sqrt{\frac{\bar{D}t}{\pi}} - 3\frac{\bar{D}t}{R^2}, \tag{5.1}$$

where F is the degree of completion of the process, Q_t is the number of sorbed ions at the time moment t, Q_∞ is the analogous value upon the completion of the process (for equilibrium conditions), R is the radius of sorbent grains and $\bar{D}$ is the coefficient of quasi-diffusion of the substance in the sorbent grain.

For high values of Fourier number $\left(\frac{\bar{D}t}{R^2} > 0.15\right)$, Eq. (5.1) becomes

$$F = 1 - \frac{6}{\pi^2} e^{-\frac{\pi^2\bar{D}t}{R^2}}. \tag{5.2}$$

This equation is the first term of an infinite series which appears in the rigorous solution of the quasi-diffusion. This equation describes the regular process of quasi-diffusion. For the low values of the Fourier number (irregular quasi-diffusion) it is necessary to use Eq. (5.1) or Boyd-Barrer approximation [105, 106] for the first term in Eq. (5.1)

$$F = 6\frac{\bar{D}t}{\pi R^2}. \tag{5.3}$$

In order to prove the limitation of the process by quasi-diffusion in the grain, it is most convenient to carry out experiments with phase contact interruption [107]. For the sorption of organic ions after this interruption and the restoration of the contact after a certain period of time, the dependence usually the shape of

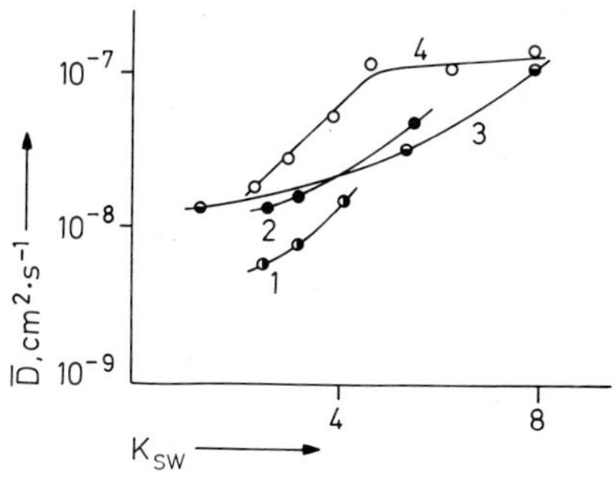

Fig. 28. Effect of degree of swelling on diffusion coefficient of streptomycin in carboxylic cation exchangers: *1*) KB-4, *2*) KFA, *3*) KB-2, *4*) KMDM-6

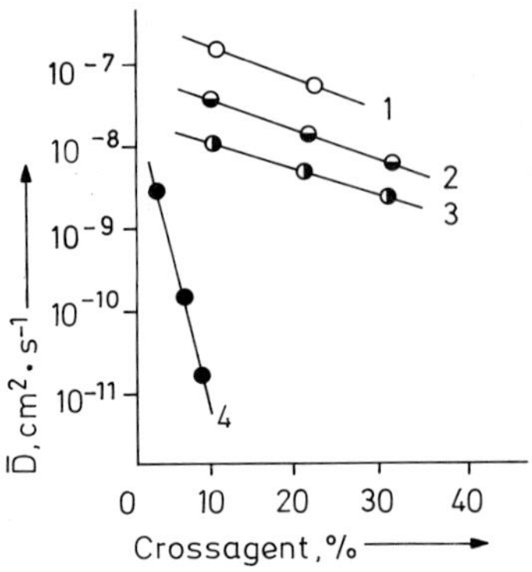

Fig. 29. Effect of the extent of crosslinking of sulfonated cation exchangers (quantity of crosslinking agent) on diffusion coefficient of tetracycline in sorbent grains: *1)* SDMDMA, *2)* SHMDMA, *3)* SEDMA, *4)* Dowex-50 W

a curve with a break, which is usually observed for mass-exchange with the participation of organic ions.

For slow diffusion of organic ions, one of the most suitable methods for the determination of quasi-diffusion in the grain is the use of the calculation of mean sorption time $\bar{t}_1$ [108]

$$\bar{t}_1 = \int_0^{\infty} t \frac{dF}{dt} dt = \int_0^1 t \, dF . \tag{5.4}$$

For a spherical grain, the following equation is valid:

$$\bar{t}_1 = \frac{R^2}{15\bar{D}} . \tag{5.5}$$

The application of the above methods of calculation has shown that the quasi-diffusion of organic counterions is profundly affected by both the amount of the crosslinking agent in a crosslinked polyelectrolyte and the method of formation of the crosslinked structure [109–112]. Fig. 28 shows the dependence of diffusion coefficients for streptomycin ions on the amount of the crosslinking

Table 14. Kinetics of quasi-diffusion of oxytetracycline ions in sulfonated cation exchangers

pH	$\bar{D}$		
	SBS-3	DOWEX 50 × 4	
2.0	$7.5 \cdot 10^{-9}$	$0.64 \cdot 10^{-9}$	counterion
11.0	$1.4 \cdot 10^{-7}$	—	coion
11.5	$2.0 \cdot 10^{-7}$	$0.96 \cdot 10^{-7}$	coion

Table 15. Quasi-diffusion of organic ions in macroporous (KU-23) and gel (SDV-T) sulfonated cation exchangers

Organic ion	$\bar{D}$	
	SDV-T	KU-23
Insulin	$1.0 \cdot 10^{-11}$	$1.1 \cdot 10^{-9}$
Tetracycline	$4.8 \cdot 10^{-8}$	$2.6 \cdot 10^{-9}$

agent in CP. The quasi-diffusion of organic ions proceeds particularly slowly in highly crosslinked gel structures. In this case, the value of $\bar{D}$ decrease considerably with increasing amount of the crosslinking agent (degree of hydration). In macroreticular polyelectrolytes, the diffusion of the same ions proceeds much faster. Moreover, the decrease in $\bar{D}$ with increasing crosslinking is not so pronounced. The data on the diffusion of tetracycline ions on macroreticular crosslinked polyelectrolytes are more significant (Fig. 29). The increase in the amount of the crosslinking agent in a KU-1 gel – CP leads to a very sharp drop in $\bar{D}$. However, in macroreticular polyelectrolytes, fast diffusion of tetracycline takes place.

The rate of the quasi-diffusion of organic ions depends on the ratio of signs of the charges of fixed ions in CP to those of diffusing ions. It follows from Table 14, on passing from counterions to coins, a drastic increase in $\bar{D}$ is observed. The use of macroporous CP does not always to the attainment of the high rate of mass-exchange for organic ions (Table 15). Insulin diffusion takes place in a macroporous CP only along channels, and a high mass-exchange rate is attained. In contrast to insulin, oxytetracycline diffuses not only along the channels but also in a densely cross-linked structure surrounding the channels. As a result, the rate of mass – exchange is lower then that of analogous processes occurring on the gel – CP. Hence, the counterions move from one functional group to another overcoming the energetic barrier of interaction between a fixed ion and a counterion, whereas coions move in the CP matrix with a very low energy expenditure.

5.2 Mass-Exchange Kinetics of Organic Ions on Composite Crosslinked Polyelectrolytes

In order to carry out some processes of heterogeneous mass-exchange on crosslinked polyelectrolytes, e.g., in the development of preparative selective low pressure chromatography of organic BAS, the kinetic permeability of macro- and heteroreticular polyelectrolytes is often insufficient. The problem of a further increase in the rate of the quasi-diffusion of organic ions in CP has been successfully solved by using composite materials. For this purpose, cellosorbents have been obtained. They are microdisperse grains of CP immobilized in a porous matrix of spherical cellulose during its formation [113–118]. Ion diffusion in these materials proceeds in pores at rates comparable to those of diffusion in liquids. The limiting

Table 16. Coefficients of quasi-diffusion of organic ions in a CS-KU-23 composite in comparison with the corresponding values for KU-23 micrograins

Organic ion	Ion exchanger	Size, μm		$\bar{D}$ ($cm^2 \cdot s^{-1}$)
		micro-grains	composite grains	
Insulin	KU-23	—	500	$3.0 \cdot 10^{-9}$
Insulin	CS-KU-23	~1	500	$3.1 \cdot 10^{-8}$
Insulin	CS-KU-23	12	500	$1.7 \cdot 10^{-8}$
cocarboxylase	KU-23	—	400	$2.1 \cdot 10^{-8}$
cocarboxylase	CS-KU-23	28	300	$4.1 \cdot 10^{-7}$
cocarboxylase	CS-KU-23	28	225	$3.2 \cdot 10^{-7}$

stage is usually diffusion in micrograins incorporated in the porous system. However, since the size of micrograins is small and the diffusion path is short, in this stage also mass-exchange proceeds at a relatively high rate. Considering cellosorbents as a biporous system, one can use the corresponding model for calculating two-stage diffusion [119]. As shown by the theoretical analysis [120], in the diffusion coefficient in transport pores D_i and that micrograins D_a can be introduced in the range of the biporous model. In this case, the effective quasi-diffusion coefficient in the composite is related to D_i and D_a by the equation:

$$\bar{D} = \frac{R_i^2}{\dfrac{R_i^2}{D_i} + \dfrac{R_a^2}{D_a}} \tag{5.6}$$

where R_i is the radius of the composite grain and R_a is the radius of the micrograin.

Furthermore, for calculating the effective coefficient of quasi-diffusion in a composite ($\bar{D}$) with the corresponding limitation of the entire process of heterogeneous mass-exchange, equations reported in Section 5.1 may be used. The high kinetic permeability of cellosorbents for large organic ions are listed in Table 16.

6 Preparative Selective High-Capacity Ion-Exchange Chromatography of Biologically Active Compounds on Crosslinked Polyelectrolytes (Biosorbents)

At present, the problem of preparative chromatography in its general form and the preparative chromatography of BAS is a widely investigated field [15, 54, 59, 121]. However, the modern approaches including the problem of efficiency and economy have been developed actively mainly in recent years and have been considered at the international symposia on liquid chromatography since 1986. A special journal on preparative chromatography began to be published in 1989.

The aim of this Section is to consider special problems of the preparative chromatography of BAS with the use of *heteroreticular*, *macroreticular* and *composite sorbents*, the structural features and some physico-chemical properties of which are considered above. As already mentioned, we call these CP groups by a general term, *Biosorbents*. Considerable sorption capacity and high selectivity of bonding BAS ions on biosorbents in combination with fast exchange rate enabled separation of substances with similar properties under the conditions of high efficiency at low pressure. One of the ideas of the development of economical preparative chromatography is the formation of broad zones of substances under the conditions of selective elution (desorption) of each zone.

In preparative selective chromatography, the formation of broad zones of the substances is determined by the formation of sharp boundaries of each zone. The formation of these sharp boundaries of substance zones in column sorption processes for systems in which the interphase transfer is limited by substance diffusion in sorbent grains [104, 122, 123] is determined by the dimensionless criterion λ:

$$\lambda = 3(1 - \alpha)\frac{\bar{D}K}{R^2V}X \tag{6.1}$$

where α is the fraction of column volume occupied by the solution, K is the distribution coefficient of substances between the sorbent and the solution, V is the rate of solution motion through the column and X is the column height.

In order to consider chromatographic processes in a more universal manner as processes in which film control of heterogeneous mass-exchange is also possible, dimensionless criteria for the conditions of formation of sharp zone boundaries may be represented by the parameter Λ [124, 125]. The evaluation of this parameter is carried out on the basis of dynamic (chromatographic) and kinetic experiments:

$$\Lambda = \frac{t_{eq} - t_0}{5\bar{t}_1} \tag{6.2}$$

where t_{eq} is the elution time of a sharp equilibrium front (or the inflection point for a symmetric curve of the frontal process), t_0 is the time of filling the free volume of the column at solvent flow rates corresponding to experimental data for t_{eq} and $\bar{t}_1$ is the mean sorption time or the first statistical moment of the kinetic curve: $\bar{t}_1 = \int_0^1 t\,dF$, where F is the degree of completion of the sorption process at a constant concentration in solution.

The sharpening and spreading of substance zones are determined by sorption conditions: the formation of a sharp boundary at $\lambda, \Lambda > 4$ and boundary spreading at $\lambda, \Lambda < 4$.

The intermediate values correspond to the approximation of the quasi-equilibrium conditions of sharpening of zone boundaries. At the same time, selective displacement in equilibrium sorption dynamics conception is carried out with the formation of a sharp rear boundary of a broad zone only if the distribution coefficient of the substance between the sorbent and the solution drastically decreases in comparison to 1. Thus, the theory of equilibrium dynamics of ion-exchange sorption processes requires the transition from $K \gg 1$ to $K \ll 1$ (K is the constant of selectivity of ion exchange) for attaining the conditions of a sharp displacement of the zone of sorbed ions. This fact should be taken into account in Eqs. (6.1) and (6.2) [54, 59].

One of the most rational means for displacing a broad zone is electrolyte desorption under the conditions of decreasing degree of ionization, i.e., when counterions are converted into dipolar ions, uncharged molecules and coions. This conversion corresponds to a sharp decrease in distribution coefficients of the desorbed substance. Hence, the displacement of equilibrium parameters at a high rate of mass-exchange is one of the methods of selective stepwise chromatography.

Selective heteroreticular biosorbents were used for direct sorption from the culture liquid of a number of enzymes with a high sorption capacity and subsequent stepwise elution (in a single-stage process) of individual components with the preparation of highly pure substances. It should be noted that the diffusion coefficients of proteins (enzymes) on biosorbents are of the order of magnitude of 10^{-9} cm^2/s. Although this value exceeds by a factor 10^2 the diffusion coefficients of proteins in conventional highly permeable gel ion exchangers, it does not make it possible to carry out the process at high flow rates on sorbents with a grain diameter of 200–400 μm, i.e. under economically advantageous conditions without additional pressure. Thus, Fig. 30 shows the elution curves for the preparative chromatographic separation of alkaline and acidic proteases from the culture liquid of Act. spheroidas in a single-stage process [126]. A similar process of direct isolation of the thrombolytic enzyme, urokinase, of sufficiently high purity for medical use has been carried out on a column with a Biocarb-T heteroreticular carboxylic cation-exchanger [127]. The enzymes were desorbed by a shift in pH farther from the isoelectric points. The distribution coefficients of the substance between the solution and the sorbent had attained tens and even hundreds. However, the relatively low rate of heterogeneous mass-exchange not only required the limitation of the flow rate in desorption to the level of 10–20 ml/cm^2/h, but

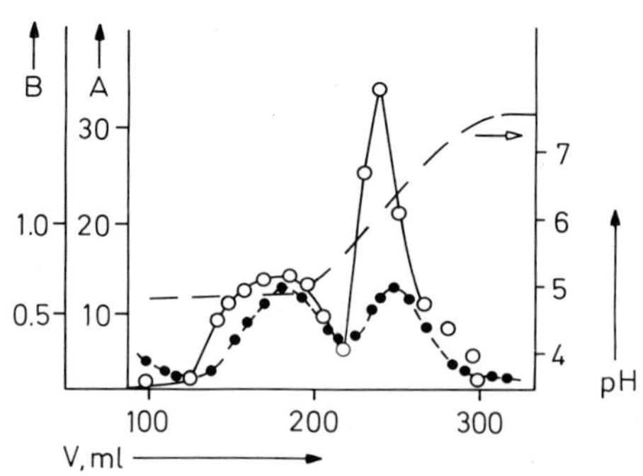

Fig. 30. Chromatographic separation of *1*) acid and *2*) neutral proteases *(Bac. Subtilis)* on Biocarb-T biosorbent. *A.* proteolytic activity units/ml —o—; *B.* protein concentration (E_{280}) —•—

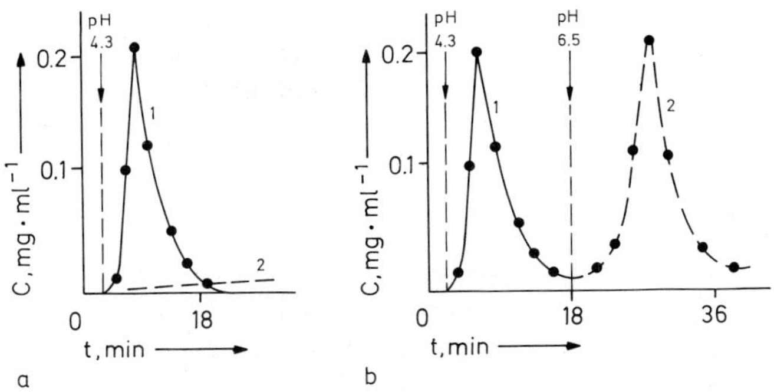

Fig. 31a, b. Elution curves of frontal desorption of thiamine mono-(TMP) and thiamine diphosphate (TDP) on a CS-KU-2 cellosorbent. **a**) *1*) TDP ($\Lambda = 11.9, \bar{t} = 7$ s), *2*) TMP ($\Lambda = 0.3, \bar{t} = 1000$ s); **b**) *1*) TDP ($\Lambda = 11.9, \bar{t} = 7$ s), *2*) TMP ($\Lambda = 3.3, \bar{t} = 56$ s)

also led to the desorption with the main substance of certain amounts of other components with isoelectric points below pH of the eluting solution in cation exchange, although many components had previously been eluted at pH < pI of the desired component.

Further improvement in preparative selective chromatography required the preparation of a new generation of biosorbents with improved kinetics of mass-exchange for organic ions. For this purpose, composite polymer sorbents, cellosorbents, were used. As already mentioned, they are microdisperse forms of selective ion exchangers immobilized in porous highly permeable matrices. When cellosorbents are used, the rate of mass-exchange increases. Accordingly, chromatography at a relatively high flow rate is possible and the process is shorter by more than one order of magnitude. A further important factor in the use of cellosorbents for preparative selective chromatography was the possiblity of using smaller shifts in pH for the desorption of the required component (Fig. 31).

Fig. 32 shows a stepwise separation of broad zones of thiamine mono- and diphosphate on a CS-KU-2 cellosorbent. The dynamic criteria (Λ and λ) predict the zones limited by sharp boundaries with a complete yield of the components at pH 4.9 for thiamine diphosphate and at pH 6.5 for thiamine monophosphate. It is noteworthy that the criterion Λ predetermines the slow motion ot the thiamine

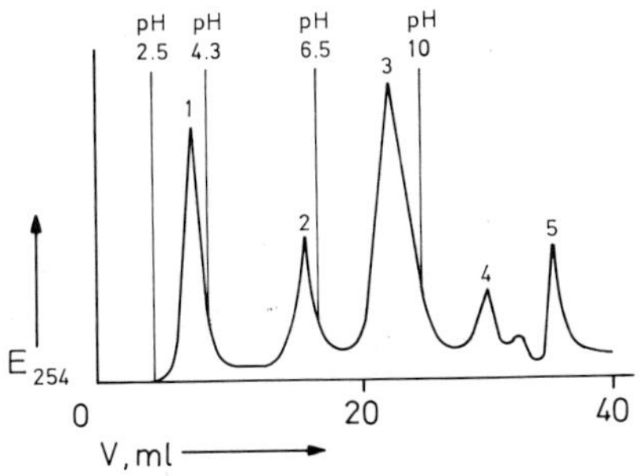

Fig. 32. Stepwise preparative desorption chromatography of phosphoric esters of thiamine on CS-KU-2 cellosorbent *1*) highly phosphoric thiamine esters, *2*) thiamine diphosphate, *3*) thiamine monophosphate, *4*) thiamine, *5*) impurities

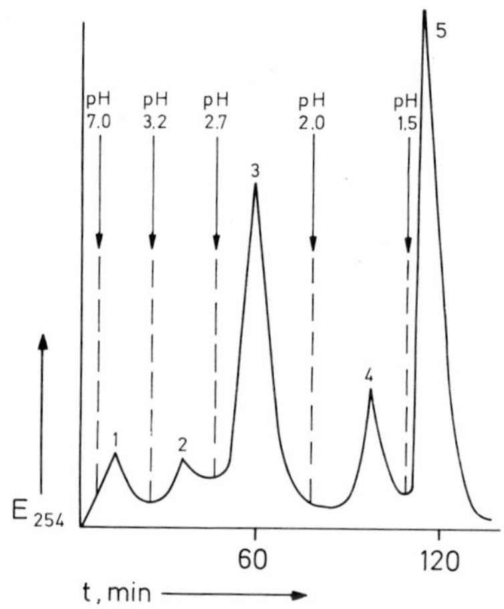

Fig. 33. Stepwise preparative desorption chromatography of phosphoric esters of adenosine on CS-AV-17 cellosorbent: *1*) adenine, *2*) adenosine, *3*) adenosine monophosphate, *4*) adenosine diphosphate, *5*) adenosine triphosphate

monophosphate zone at pH 4.3. The formation of several zones of substances at a high concentration is favored by high mass exchange rates. This is demonstrated by low values of mean sorption times ($\bar{t}_1$) for the conditions of displacement of the components with the formation of sharp zone boundaries.

These ideas and methods of preparative selective up-scale chromatography suggest that the use of new types of biosorbents and, in particular cellosorbents, and the application of theoretically based conditions for stepwise desorption of the components is an important new approach to preparative chromatography.

7 References

1. Dušek K (1965) J Polym Sci (B) 3: 209
2. Dušek K (1967) J Polym Sci (C) 16: 1289
3. Dušek K (1971) in: Shompff AJ, Newmann S (eds) Polymer Networks. Structure and Mechanical Properties. Plenum Press, New York, p 245
4. Jacobelli H, Bartholin M, Guyet A (1979) J Appl Polym Sci 23: 927
5. Dušek K (1982) in: Development in Polymerization. 3. Haward RN (ed) Appl Sci Publ Barking, p 143
6. Millar JR, Smith DG, Marr WF, Kressman TRE (1963) J Chem Soc 218
7. Millar JR, Smith DG, Kressman TRE (1965) J Chem Soc 304
8. Tager AA, Tsilipotkina MV, Makovskaya EB et al. (1971) Vysokomol Soedin (A) 13: 370
9. Tager AA, Tsilipotkina MV, Makovskaya EB et al. (1968) Vysokomol Soedin (A) 10: 1065
10. Irzhak VJ, Rosenberg BA, Enikolopian NS (1979) Crosslinked polymers (in Russian), Nauka
11. Tager AA, Tsilipotkina MV (1978) Usp Khim 47: 152
12. Dušek K, Galina H, Mikeš J (1980) Polym Bull 3: 19
13. Katchalsky A, Lifson S (1953) J Polym Sci 11: 409
14. Galina H, Kolarz BN (1979) J Appl Polym Sci 24: 891, 901
15. Samsonov GV (1986) Ion Exchange and Preparative Chromatography of Biologically Active Molecules. Plenum Press corp consult Bureau, New York
16. Selezneva AA, Dubinina NI, Babenko GA, Luknitskaya OF, Samsonov GV (1975) Kolloid Zhurn 37: 1138

17. Kuznetsova NN, Rozhetskaya KM, Moskvichev BV, Samsonov GV et al. (1976) Vysokomol Soedin (A) 18: 355
18. Samsonov GV, Kuznetsova NN, Yurchenko VS, Papukova KP et al. (1979) Vysokomol Soedin (B) 21: 244
19. Samsonov GV, Selezneva AA (1981) Khim Farm Zhur 7: 77
20. Kuznetsova NP, Mishaeva RN, Kuznetsova NN, Rozhetskaya KM, Samsonov GV (1980) Vysokomol Soedin (A) 22: 874
21. Rozhetskaya KM, Kalinina NA, Papukova KP, Kallistov OV, Samsonov GV (1989) Vysokomol Soedin (A) 31: 2532
22. Flory JP (1953) Principles of Polymer chemistry. Cornell Unit Press, Ithaca
23. Yurchenko VS (1977) Physico-chemical Properties of Macroreticular Polyelectrolytes in Connection with Specific Features of Sorption of Organic Substances. (In Russian) Dissertation, Institute of Macromol Compounds, Acad Sci USSR, Leningrad
24. Pisarev OA, Muravieva TD, Samsonov GV (1986) Vysokomol Soedin (B) 28: 262
25. Pisarev OA, Dobrodumov AV, Muravieva TD, Denisov VM, Koltsov AI, Samsonov GV (1986) Vysokomol Soedin (B) 28: 362
26. Gavrilova NN, Pirogov VS, Morozova AD, Nadezhin YuS, Kuznetsova NN, Samsonov GV (1981) Zhur Prikl Khim 54: 1190
27. Ezhova NM, Chernova IA, Pogodina TE, Samsonov GV (1984) in: Samsonov GV (ed) Ion Exchange and Chromatography, Nauka, p 38
28. Chernova IA, Samsonov GV (1979) Vysokomol Soedin (A) 21: 1608
29. Colombo VF, Spath PV (1981) Int J Scan Electron Microsc Relat Techn and Appl, p 515
30. Kuznetsova NP, Mishaeva RN, Gudkin LR, Anufrieva EV, Pautov VD, Samsonov GV (1977) Vysokomol Soedin (A) 19: 107
31. Kuznetsova NP, Mishaeva RN, Gudkin LR, Kuznetsova NN, Muravieva TD, Papukova KP, Rozhetskaya KM, Samsonov GV (1978) Vysokomol Soedin (A) 20: 629
32. Anufrieva EV, Kuznetsova NP, Krakoviak MG, Mishaeva RN, Pautov VD, Semisotnov GV, Sheveleva TV (1977) Vysokomol Soedin (A) 19: 102
33. Anufrieva EV, Pautov VD, Kuznetsova NP, Mishaeva RN (1981) Vysokomol Soedin (B) 23: 557
34. Kuznetsova NP, Mishaeva RN, Kuznetsova NN, Rozhetskaya KM, Samsonov GV (1980) Vysokomol Soedin (B) 22: 874
35. Frolov VI, Kuznetsova NP, Gudkin LR, Mishaeva RN (1977) Vysokomol Soedin (A) 19: 984
36. Katchalsky A, Michaeli J (1955) J Polym Sci 15: 69
37. Gregor HP, Frederick M (1957) J Polym Sci 23: 451
38. Oosawa F (1971) Polyelectrolytes, M Dekker, New York
39. Fichtner F, Schonert H (1977) Colloid Polym Sci 255: 230
40. Anufrieva EV, Volkenshtein MV, Krakoviak MG, Sheveleva TV (1968) Dokl Akad Sci USSR 182: 361
41. Nekrasova TN, Anufrieva EV, Eljashevich AM, Ptitsin OB (1965) Vysokomol Soedin 7: 913
42. Leyte JC, Mandel M (1964) J Polym Sci (A) 2: 1879
43. Irie M (1984) Makromol Chem, Rapid Commun 5: 413
44. Anufrieva EV, Gotlib YuYa (1981) in: Cantow HJ, Dall Asta G, Dušek K (eds) Investigation of Polymers in Solution by Polarized Luminescence, Springer, Berlin, Heidelberg, New York (Adv Polym Sci, 40)
45. Anufrieva EV (1982) Pure Appl Chem 54: 533
46. Radziavichius KI, Shataeva LK, Samsonov GV, Zhukova NT, Zorina AI, Laskorin BN (1982) Vysokomol Soedin (A) 24: 1066
47. Chernova IA, Pogodina TE, Shataeva LK, Samsonov GV (1980) Vysokomol Soedin (A) 22: 2403
48. Krakoviak MG, Anufrieva EV, Volkenshtein MV, Ananieva TD, Gotlib YuYa, Gromova RA, Kozel SP, Lashkov GI, Lushchik VB, Pautov VD, Skorokhodov SS, Sheveleva TV (1975) Dokl Akad Sci USSR 224: 873

49. Mishaeva RN, Kuznetsova NP, Samsonov GV, Gudkin LR, Papukova KP, Rozhetskaya KM (1988) Vysokomol Soedin (A) 30: 1707
50. Bonner OD, Holland VF, Smith LL (1956) J Phys Chem 60: 1102
51. Hogfeldt E, Ekedahl F, Sillen LC (1950) Acta Chem Scand 4: 828
52. Soldatov VS, Pokrovskaya AI, Martsinkevich RV (1967) Zhur Fiz Khim 41: 1098
53. Samsonov GV, Pasechnik VA (1969) Usp Khim 38: 1257
54. Samsonov GV, Trostianskaya EB, Elkin GE (1969) Ion Exchange. Sorption of Organic Substances. (In Russian) Nauka
55. Kressman TRE, Kitchener JA (1949) J Chem Soc 1280
56. Moskvichev BV, Samsonov GV (1967) Izv Akad Nauk USSR ser Khim 742
57. Moskvichev BV, Pasechnik VA, Musabekov KB, Chokina BSh, Smirnova PM, Samsonov GV (1970) Zhur Fiz Khim 44: 2589
58. Samsonov GV (1974) Pure Appl Chem 38: 151
59. Samsonov GV (1960) Sorption and Chromatography of Antibiotics (in Russian). Izd Akad Nauk USSR, Moskow
60. Samsanov GV (ed) (1968) Selectivity of Ion-Exchange Sorption of Antibiotics. (in Russian) Transaction of Leningrad Chemico-Pharmaceutical Institute, Leningrad
61. Bakaeva RM, Samsonov GV (1970) Antibiotiki (in Russian) 15: 985
62. Lumry R, Rajender Sh (1970) Biopolymers 9: 1125
63. Shataeva LK, Kuznetsova NN, Elkin GE (1979) Carboxylic Cation-Exchangers in Biology. (in Russian) Nauka
64. Dmitrenko LV, Morozova AD, Samsonov GV (1972) Izv Akad Nauk, ser Khim 7: 1563
65. Pirogov VS, Nemtsova NN, Borisova VA, Samsonov GV (1981) Zhur Fiz Khim 55: 2937
66. Landau MA (1981) Molecular Mechanisms of Effect of Physicologically Active Compounds. (in Russian) Nauka
67. Laurence DJR (1972) in: Carroll B (ed) Physical Methods in Macromolecular Chemistry, M Dekker, New York (Vol 2)
68. Kuznetsova NP, Samsonov GV, Mishaeva RN (1989) Vysokomol Soedin (A) 31: 723
69. Kabanov VA, Zezin AB, Mustafaev MI, Kasaikin VA (1980) in: Gaethals EJ (ed) Polymeric Amines and Ammonium Salts. Pergamon Press, Oxford, p 173
70. Kabanov VA, Mustafaev MI (1981) Vysokomol Soedin (A) 23: 255
71. Morawetz H (1965) Macromolecules in Solutions. Interscience — J Wiley, New York, p 92
72. Pautov VD, Kuznetsova NP, Mishaeva RN, Anufrieva EV (1983) Vysokomol Soedin (A) 25: 1599
73. Anufrieva EV, Kuznetsova NP, Pautov VD (1989) Vysokomol Soedin (B) 31: 355
74. Poltorak OM, Chukhrai ES, Priakhin AN (1979) in: Berezin IV, Martinek K (eds), Progress in Bioorganic Catalysis (in Russian), Ed Moscow University, Moscow, p 57
75. Kamyshnyi AL (1981) Zhur Fiz Khim 55: 562
76. Wyman JJr (1964) in: Anfinsen CB, Edsall JT, Richards FM (eds) Advances of Protein Chemistry, Acad Press, New York (vol 19), p 224
77. Pisarev OA, Kuznetsova NP, Mishaeva RN, Samsonov GV (1985) Vysokomol Soedin (B) 27: 261
78. Lamry R, Biltonen R (1969) in: Timasheff SN, Fasman GD (eds) Structure and Stability of Biological Macromolecules. M Dekker, New York, p 7
79. Nemtsova NN, Pirogov VS, Samsonov GV (1984) in: Samsonov GV (ed) Ion-Exchange and Chromatography. Nauka, p 76
80. Nemtsova NN, Borisova VA, Pirogov VS, Samsonov GV (1984) Zhur Fiz Khim 58: 1269
81. Naumova LV, Vorobieva VYa, Samsonov GV (1980) Zhur Prikl Khim 53: 764
82. Plate NA, Vasiliev AE (1986) Physiologically Active Polymers. (in Russian) Khimiya, Moscow
83. Chang TMS (1972) Artificial Cells. C Thomas Publ Springfield, Illinois
84. Chang TMS (1976) in: Mosbach K (ed) Methods in Enzymology. Acad Press, New York (vol 44), p 201
85. Kazanskaya NF, Aisina RB (1978) in: Itogi Nauki i Techniki, ser Chemistry and Technology of Macromolecular Compounds. (in Russian) VINITI, Moscow, vol 12, p 115

86. Gregoriadis G (1976) in: Mosbach K (ed) Methods in Enzymology. Acad Press, New York (vol 44) p 218
87. Margel S (1984) J Polym Sci 22: 3521
88. Samsonov GV (1979) Vysokomol Soedin (A) 21: 723
89. Orlievskaya OV, Morozova EN, Ponomareva RB, Samsonov GV (1976) Kolloid Zhur 38: 1182
90. Ivanova GP, Mirgorodskaya OA, Selezneva AA, Moskvichev BV, Samsonov GV (1976) Prikl Biokhim Mikrobiol 12: 33
91. Timkovskaya AF, Mirgorodskaya OA, Selezneva AA, Moskvichev BV, Samsonov GV (1976) Prikl Biokhim Mikrobiol 12: 886
92. Kiseleva EM, Moskvichev BV, Momot NN, Mirgorodskaya OA, Samsonov GV (1977) Prikl Biokhim Mikrobiol 13: 721
93. Yurchenko VS, Kostareva IA, Ponomareva RB, Samsonov GV (1983) Prikl Biokhim Mikrobiol 19: 528
94. Ponomareva RB, Kostareva IA, Orlievskaya OV, Yurchenko VC, Samsonov GV (1983) Vysokomol Soedin (B) 25: 415
95. Samsonov GV, Kuznetsova NP, Gudkin LR, Mishaeva RN (1981) Dokl Akad Nauk USSR 260: 1486
96. Antonini E, Brunori M (1971) in: Frontiers in Biology. Amsterdam (vol 21)
97. Antonini E, Rossi-Bernardi L, Chiancone E (1981) in: Colowick SP, Kaplan NO (eds) Methods in Enzymology. Acad Press, New York
98. Damaschun G, Damaschun H, Gedicke C et al. (1975) Acta Biol Med Germ 34: 391
99. Nurith S, Hans A (1980) Biochem Biophys Res Communs 95: 1105
100. Tenforde T (1970) in: Advances in Biological and Medical Physics. Acad Press, New York (vol 13), p 43
101. Libinson GS, Savitskaya EM, Bruns BP (1963) Zhur Fiz Khim 37: 420
102. Barrer RM, Bartholomeu RF (1963) J Phys Chem Solids 24: 309
103. Znamenskii YuP, Kasperovich AI, Bychkov NV (1968) Zhur Fiz Khim 42: 2017
104. Samsonov GV, Elkin GE (1985) in: Marinsky JA, Marcus J (eds) Ion Exchange and Solvent Extraction, M Dekker, New York, p 211
105. Barrer RM (1941) Diffusion In and Through Solids. Cambridge Univ. Press, New York
106. Boyd GE, Adamson AW, Myers LS (1947) J Amer Chem Soc 69: 2836
107. Kressman TRE, Kitchener JA (1949) Disc Farad Soc 7: 90
108. Tunitsky NN, Kaminsky VA, Timashev SF (1972) Methods of Physico-Chemical Kinetics (in Russian) Khimiya, Moscow
109. Dinaburg VA, Genender KM, Pasechnik VA, Yurchenko VS, Elkin GE, Belaya SF, Samsonov GV (1968) Zhur Prikl Khim 41: 891
110. Belaya SF, Elkin GE, Samsonov GV (1968) Antibiotiki 13: 737
111. Genedi AM, Samsonov GV (1969) Kolloid Zhur 31: 674
112. Genedi AM, Tevlina AS, Samsonov GV (1969) Antibiotiki 14: 99
113. Samsonov GV, Melenevsky AT, Demin AA, Dubinina NI, Tishchenko GA, Papukova KP, Elkin GE, Pirogov VS, Chizhova EB (1984) Ion Exchange and Chromatography, Samsonov GV (ed) Nauka, p 100
114. Yaskovich TA, Elkin GE, Samsonov GV, Papukova KP (1984) Ion Exchange and Chromatography, Samsonov GV (ed) Nauka, p 113
115. Demin AA, Melenevsky AT, Dubinina NI, Papukova KP, Pirogov VS, Samsonov GV (1984) Zhur Prikl Khim 57: 2212
116. Chizhova EB, Melenevsky AT, Pirogov VS, Samsonov GV (1985) Zhur Prikl Khim 58: 1091
117. Melenevsky AT, Nemtsova NN, Papukova KP, Pirogov VS, Samsonov GV (1985) Zhur Prikl Khim 58: 1257
118. Melenevsky AT, Demin AA, Pirogov VS, Papukova KP, Dubinina NI, Samsonov GV (1986) Zhur Prikl Khim 59: 610
119. Zolotarev PP, Dubinin MM (1973) Dokl Akad Nauk USSR 210: 136
120. Nikolaev NI, Zolotarev PP, Popkov YuM, Ulin VI (1981) Theory and Practice of Sorption Processes. (in Russian) Voronezh University, Voronezh 14: 12

121. Wankat PhC (1986) Large-Scale Adsorption and Chromatography, CRC Press Inc Baca Raton, Florida
122. Samsonov GV, Elkin GE, Lebedev YuYa, Momot NN (1975) in: Samsonov GV (ed) Ion exchangers and ion exchange. (in Russian) Nauka, p 98
123. Samsonov GV, Elkin GE, Melenevsky AT, Samsonov GV (ed) (1984) Ion Exchange and Chromatography, Nauka, p 94
124. Elkin GE, Melenevsky AT, Chizhova EB, Samsonov GV (1982) Izv Akad Nauk SSSR, ser Khim 4: 946
125. Chizhova EB, Elkin GE, Melenevsky AT, Samsonov GV (1983) Izv Akad Nauk SSSR, ser Khim 12: 2749
126. Pannel KE, Kalyula KhYa, Sakaluskaite NYu, Letunova EV, Tikhomirova AS, Shataeva LK, Samsonov GV (1975) Prikl Biokhim Mikrobiol 11: 598
127. Koltsova SV, Shataeva LK, Sukhareva TF, Byniaeva NA, Fedorova ZD (1981) Voprosy Med. Khim 5: 623

Editor: K. Dušek
Received May 13, 1991

Photoinduced Electron Transfer in Amphiphilic Polyelectrolyte Systems

Yotaro Morishima
Department of Macromolecular Science, Faculty of Science, Osaka University, Toyonaka, Osaka 560, Japan

This review article attempts to summarize and discuss recent developments in the studies of photoinduced electron transfer in functionalized polyelectrolyte systems. The rates of photoinduced forward and thermal back electron transfers are dramatically changed when photoactive chromophores are incorporated into polyelectrolytes by covalent bonding. The origins of such changes are discussed in terms of the interfacial electrostatic potential on the molecular surface of the polyelectrolyte as well as the microphase structure formed by amphiphilic polyelectrolytes. The promise of tailored amphiphilic polyelectrolytes for designing efficient photoinduced charge separation systems is also discussed.

Advances in Polymer Science, Vol. 104

1 Introduction

The primary process in natural photosynthetic systems involves a fast photoinduced electron transfer (ET) and a subsequent efficient charge separation. Molecular understanding of the factors responsible for such critical functions in the biological systems and design of relevant model systems have attracted many researchers for decades. A goal of these studies is to the development of efficient photodriven charge separation systems useful for applications as artificial photon energy conversion (into electronic energy or chemical potential), information processing, and organic photoimaging systems.

The transfer of an electron from a photoexcited donor molecule (D) to an acceptor molecule (A) to generate a highly reactive radical ion pair is the most fundamental photochemical reaction, and it can be generally expressed as

$$D^* + A \rightarrow D^+ + A^- \tag{1}$$

The back ET reaction to regenerate the original ground-state D and A is thermodynamically favorable and is very likely to occur spontaneously. Therefore, it is critically important to suppress the back ET in order to utilize the photogenerated radical ion pair for subsequent useful reactions.

It is now believed from studies on the natural photosynthetic systems that microenvironments for the photoinduced ET reaction play an important role in the suppression of the back ET [1–3]. As such reaction environments, molecular assembly systems such as micelles [4], liposomes [5], microemulsions [6–8] and colloids [9] have been extensively investigated. In them, the presence of microscopically heterogeneous phases and interfacial electrostatic potential is the key to the ET rate control.

Functionalized polyelectrolytes are promising candidates for photoinduced ET reaction systems. In recent years, much attention has been focused on modifying the photophysical and photochemical processes by use of polyelectrolyte systems, because dramatic effects are often brought about by the interfacial electrostatic potential and/or the existence of microphase structures in such systems [10, 11]. A characteristic feature of polymers as reaction media, in general, lies in the potential that they make a wider variety of molecular designs possible than the conventional organized molecular assemblies such as surfactant micelles and vesicles. From a practical point of view, polymer systems have a potential advantage in that polymers per se can form film and may be assembled into a variety of devices and systems with ease.

The polyelectrolyte "catalysis" of chemical reactions involving ionic species has been the subject of extensive investigations since the pioneering studies of Morawetz et al. [12] and Ise et al. [13–17]. The "catalytic effect" or the ability of polyelectrolytes to enhance or retard reaction rates is mainly due to concentration or exclusion of either or both of the ionic reactants by the polyions added to the reaction systems. For example, the chemical reaction between ionic species carrying the same charge is enhanced in the presence of polyions carrying the opposite charge. This enhancement can be attributed to an increase in the local concentration of the ionic species around the polyions by electrostatic attraction. The ionic reactants are distributed in a variety of ways between the polyelectrolyte molecular surface and the bulk phase, depending on the strength of electrostatic interaction. Therefore, the system is essentially dynamic in nature.

Meisel et al. [18–20] were the first to investigate how the addition of a polyelectrolyte affects photoinduced ET reactions. They found that charge separation was enhanced as a result of the retardation of the back ET when poly(vinyl sulfate) was added to an aqueous reaction system consisting of tris(2,2′-bipyridine)ruthenium(II) chloride (cationic photoactive chromophore) and neutral electron acceptors [21]. More recently, Sassoon and Rabani [22] observed that the addition of polybrene (a polycation) had a significant effect on separating the photoinduced ET products in an aqueous solution containing *cis*-dicyano-bis(2,2′-bipyridine)ruthenium(II) (photoactive donor) and potassium hexacyano-ferrate(III) (acceptor). These findings are ascribable to the electrostatic potential of the added polyelectrolytes.

2 General Considerations of Functionalized Polyelectrolytes

By covalently attaching reactive groups to a polyelectrolyte main chain the uncertainty as to the location of the associated reactive groups can be eliminated. The location at which the reactive groups experience the macromolecular environment critically controls the reaction rate. If a reactive group is covalently bonded to a macromolecular surface, its reactivity would be markedly influenced by interfacial effects at the boundary between the polymer skeleton and the water phase. Those effects may vary with such factors as local electrostatic potential, local polarity, local hydrophobicity, and local viscosity. The values of these local parameters should be different from those in the bulk phase.

The polyelectrolyte covalently functionalized with reactive groups may be viewed as an enzyme-like functional polymer or as a "molecular reaction system" in the sense that it has both reactive centers and reaction rate-controlling microenvironments bound together on the same macromolecule.

The modification of reaction rates by such a functionalized polyelectrolyte system was first made in 1959 by Ladenheim and Morawetz [23] in the Menshutkin

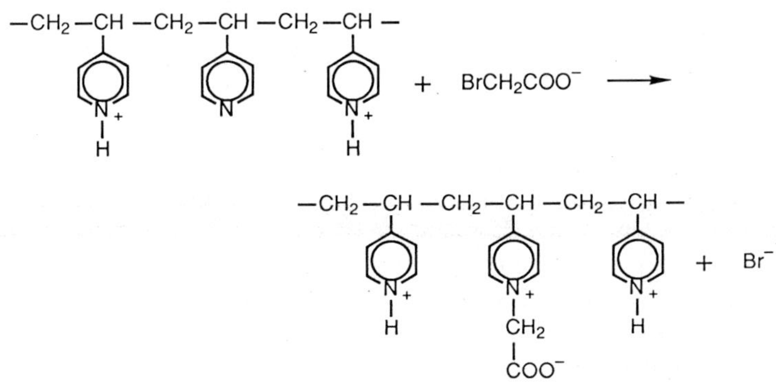

(2)

reaction of partially ionized poly(4-vinylpyridine) with β-bromoacetate. In this reaction, the partially protonated polycation electrostatically attracts the anionic carboxylate reagents to facilitate the nucleophilic attack of the unprotonated pyridine base units to the reagents. Thus, the rate of the quaternization reaction is considerably accelerated. Such a rate enhancement was not observed for β-bromoacetamide which is uncharged [23].

Ladenheim and Morawetz [23] also showed that the reactivity of the carboxylate units in partially ionized poly(methacrylic acid) (PMA) toward $BrCH_2COO^-$ in the bromine displacement reaction was greatly diminished, while the reaction proceeded at an appreciable rate with uncharged β-bromoacetamide [23]. This inhibition of the reaction of the polyanion with a small anionic reagent can be attributed to the electrostatic repulsion between the polymer and the reagent.

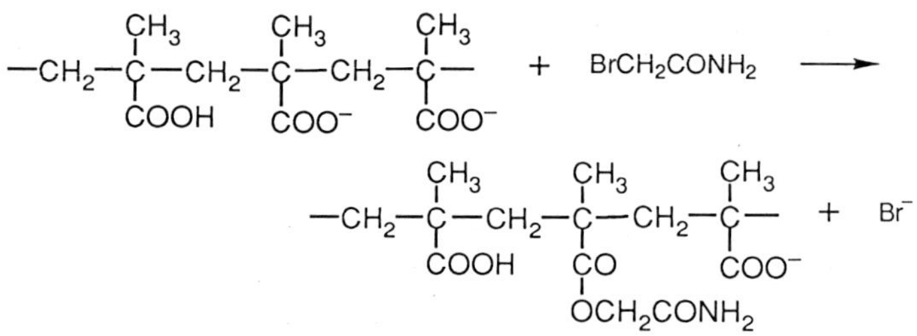

(3)

Photoresponsive polyelectrolytes tethered with a photochemical functional group were first reported in 1964 by Lovrien and Waddington [24] who prepared copolymers of *N*-azobenzeneacrylamide and acrylic or methacrylic acid (**1**).

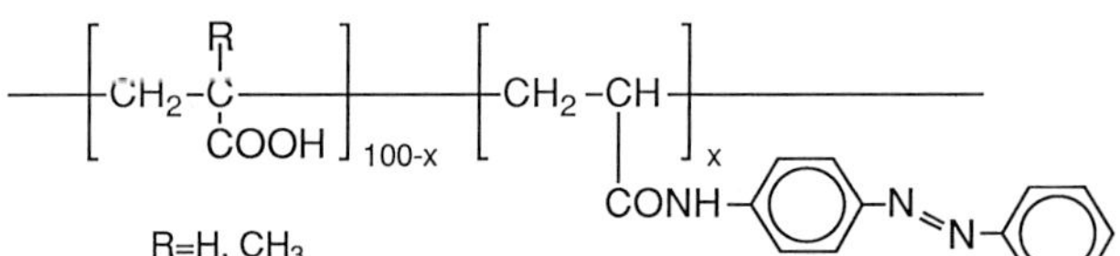

1

Irradiation with UV light isomerized the azobenzene units from the *trans* to the *cis* form, while the reverse isomerization occurred thermally in the dark. The *cis* to *trans* conversion is catalyzed by both protons and hydroxyl ions. Hence, the catalyzed dark process for tethered azobenzene is greatly modified in comparison with that for free azobenzene. For the tethered azobenzene, beginning at pH 6, the *cis* to *trans* return rate sharply decreased with increasing pH up to 10, whereas the rate for free azobenzene rapidly increased in the same pH range owing to OH^- catalysis. These observations can be explained by the electrostatic repulsion which lowers the local OH^- concentration on the polyion surface below that in the bulk aqueous phase.

Besides the electrostatic potential effect on reactivity, functionalized polyelectrolytes have a variety of interesting features worthy of study. If a polyelectrolyte is covalently modified with highly hydrophobic functional groups, it provides an unusual opportunity to study the chemical reactions of normally otherwise water insoluble functional groups in aqueous solution. Furthermore, a structural organization via hydrophobic interactions may occur in aqueous solution [25–31], which is of general scientific importance and is worth studying for its own sake.

The microphase structure of amphiphilic polyelectrolytes depends on the tendency for the hydrophobic groups in the polymer to self-aggregate intramolecularly. The attractive hydrophobic interaction between the hydrophobic groups competes with the repulsive electrostatic interaction between the charged segments in the polymer chain. If the former dominates the latter, the hydrophobic groups will form a hydrophobic microdomain whose periphery is surrounded by the charged segments. Thus, stabilization of a micelle-like microphase structure by the charged outer layer exposed to the water phase will occur.

In the following chapters, we will concern ourselves with the nature of the interfacial microenvironments of some polyelectrolytes whose functionality controls photoinduced ET. Emphasis will be placed on the local electrostatic potential and also on the microphase structure of some amphiphilic polyelectrolytes in aqueous solution.

3 Surface Potential of Polyelectrolytes

3.1 Calculation of Electrostatic Potential by the Poisson-Boltzmann Equation

The reactivity modification or the reaction rate control of functional groups covalently bound to a polyelectrolyte is critically dependent on the strength of the electrostatic potential at the boundary between the polymer skeleton and the water phase ("molecular surface"). This dependence is due to the covalent bonding of the functional groups which fixes the reaction sites to the molecular surface of the polyelectrolyte. Thus, the surface potential of the polyion plays a decisive role in the quantitative interpretation of the reactivity modification on the molecular surface.

This potential reflects itself in the titration curves of weak polyacids such as poly(acrylic acid) and poly(methacrylic acid) [32]. Apparent dissociation constants of such polyacids change with the dissociation degree of the polyacid because the work to remove a proton from the acid site into the bulk water phase depends on the surface potential of the polyelectrolyte.

The polyelectrolyte chain is often assumed to be a rigid cylinder (at least locally) with a uniform surface charge distribution [33–36]. On the basis of this assumption the non-linearized Poisson-Boltzmann (PB) equation can be used to calculate how the electrostatic potential φ varies with the distance r from the cylinder axis

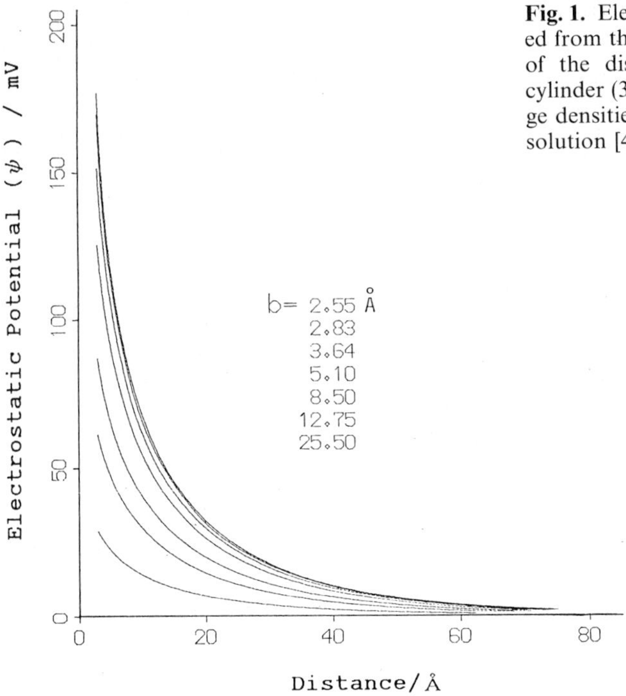

Fig. 1. Electrostatic potential calculated from the PB equation as a function of the distance from the axis of a cylinder (3 Å radius) with various charge densities in 0.016-M NaCl aqueous solution [42]

[37–39]. The PB equation in cylinder symmetry is given by

$$\frac{1}{x}\frac{d}{dx}\left(x\frac{dy}{dx}\right) = \sinh y \tag{4}$$

Here, x and y are the dimensionless distance and potential defined by $x = \varkappa r$ and $y = e\varphi/kT$, respectively, where e is the elementary charge, T the absolute temperature, k the Boltzmann constant, and $\varkappa$ the Debye screening parameter defined by $\varkappa = (8\pi ne^2/\varepsilon kT)^{1/2}$.

Figure 1 illustrates calculated potential-distance curves for polyions with different linear charge densities. Here, the charge density is expressed in terms of the average axial spacing (b) between charged groups on the polyion. The value of b = 2.55 Å is commonly used for vinyl polymers carrying one ionic charge on each monomeric residue. As can be seen from Fig. 1, for the polyion with a high charge density the electrostatic potential rapidly decreases with increasing distance from the cylinder surface.

3.2 Estimation of Electrostatic Potential by a Polymer-Bound Probe

Another approach to the estimation of the surface potential is to incorporate the moiety of an optical probe into the polyelectrolyte through covalent bonding.

A merocyanine dye, 1-ethyl-4-(2-(4-hydroxyphenyl)ethenyl)pyridinium bromide (M-Mc, **2**), exhibits a large spectral change according to the acid-base equilibrium [40, 41]. The equilibrium is affected by the local electrostatic potential and the polarity of the microenvironment around the dye. Hence, this dye is useful as a sensitive optical probe for the interfacial potential and polarity when it is covalently attached to the polyelectrolyte backbone.

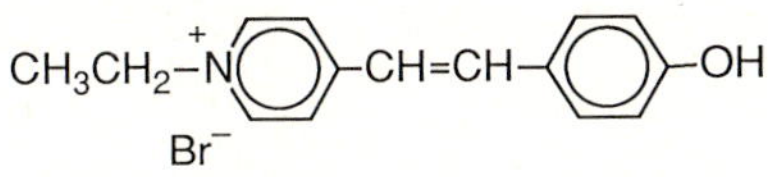

M-Mc, **2**

Morishima et al. [42] prepared polyanions (A-x, **3**) of various charge densities tagged with the merocyanine dye (Mc) by terpolymerization of acrylic acid (AA), acrylamide (AAm), and a small mole fraction of 1-(2-methacryloyloxyethyl)-4-(2-(4-hydroxyphenyl)ethenyl)pyridinium bromide (MA-Mc). Since these polyanions carry only 1 mol% of the MA-Mc units, they can practically be treated as copolymers of AA and AAm with a wide range of composition.

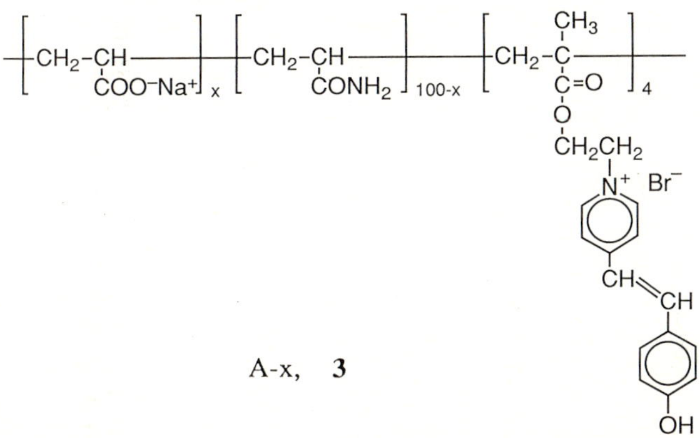

A-x, **3**

According to Manning's theory [43–45], a polyion is characterized by a dimensionless charge density parameter ξ which is defined by

$$\xi = e^2/\varepsilon kTb \tag{5}$$

where ε is the dielectric constant of water ($\varepsilon = 78$) and b the average axial spacing between charged groups on the polyion. The value of $\xi = 2.81$ is commonly used for fully extended vinylic polyions in aqueous solution at room temperature.

It is reasonable to consider that the electrostatic repulsion between the fixed line charges on the polyion tends to straighten the flexible vinylic main chain, and

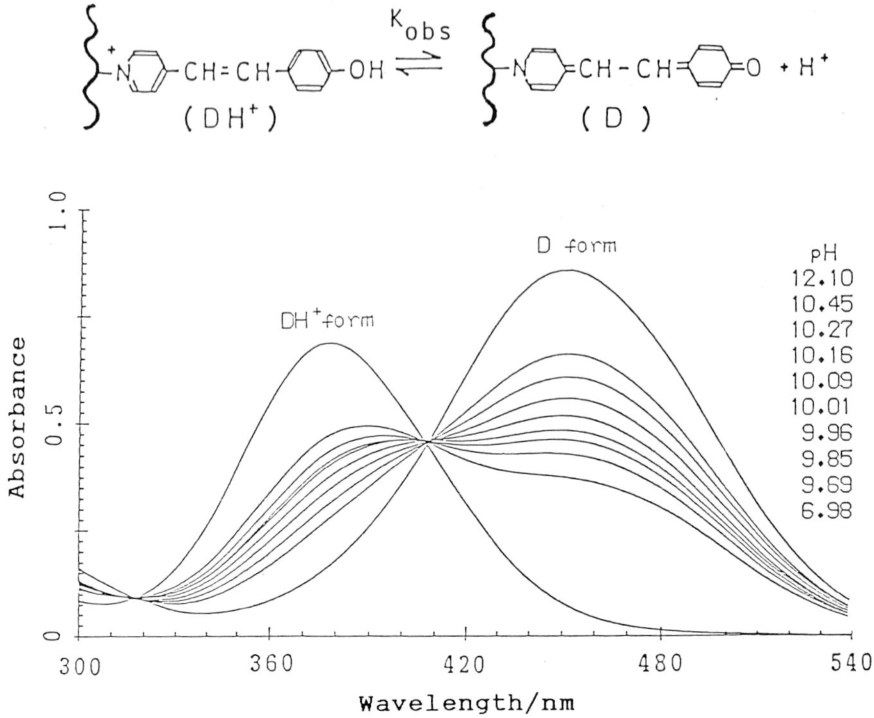

Fig. 2. A representative example of spectroscopic pH titration for A-100 (**3** with x = 100) [42]

thus allows the chain to behave at least locally as a rigid cylinder [33–36]. The ξ value for the polyion (**3**) ranges from 0.28 to 2.81, depending on the composition of the AA and AAm units, and hence 25.5 Å < b < 2.55 Å.

The acid-base equilibrium constant for the Mc residue can be determined by spectroscopic pH titration. An example for the titration is shown in Fig. 2. The electrostatic potential φ can be related to the difference between the apparent pK on the charged surface (pK_{obs}) and that on an intrinsic neutral surface (pK^i) by

$$pK_{obs} - pK^i = -F\varphi/2.3\,RT \qquad (6)$$

where F is the Faraday constant and R the universal gas constant [46, 47]. Actually, the pK^i value is not directly measurable, but Fernàndez and Fromherz [48] have shown that it can be estimated indirectly by mimicking the interfacial microenvironments by organic solvent/water mixtures. The procedure has been described in detail in the literature [48].

The merocyanine dye mentioned above shows solvatochromism, which means that the absorption band maximum of the quinoid form (D form) is sensitive to solvent polarity [40, 41]. In Fig. 3, the absorption maximum of the solvatochromic band for M-Mc (a low molecular weight merocyanine analog) is plotted against the dielectric constant of 1,4-dioxane/water mixtures [42]. With the relationship

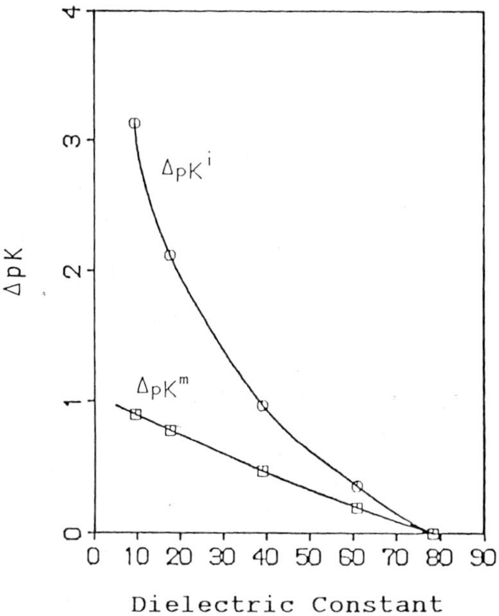

Fig. 3. Plots of ΔpK^m and ΔpK^i against the dielectric constant of 1,4-dioxane/water mixtures: $\Delta pK^m = pK^m - pK^w$, $\Delta pK^i = pK^i - pK^w$, and $pK^w = 8.52$ [42]

shown, the effective local dielectric constant ε_{eff} can be estimated. Then, the pK^i values are readily obtained by knowing the ε_{eff} values from Fig. 3 and allow φ to be computed from Eq. 6.

The values of φ thus calculated for the polyanions (**3**) having various charge densities are listed in Table 1. It should be noted that all the carboxylic acid groups in the polyions are assumed to be almost fully dissociated in the pH region where the spectroscopic pH titration of the Mc residue was performed.

Table 1. Electrostatic potentials calculated from pK_{obs} values for A − x (**3**) with various charge densities [42]

x in A − x	b^a/Å	ξ^b	$\varepsilon_{eff}{}^c$	ξ'^d	pK_{obs}	pK^i	φ/mV
100	2.55	2.8	71	3.1	9.96	8.57	82.1
90	2.83	2.5	70	2.8	9.89	8.58	77.5
70	3.64	2.0	67	2.3	9.74	8.61	66.8
50	5.10	1.4	63	1.7	9.63	8.65	58.0
30	8.50	0.84	61	1.1	9.43	8.67	45.0
20	12.8	0.56	59	0.75	9.30	8.68	36.7
10	25.5	0.28	60	0.36	9.16	8.69	29.0

[a] Average axial spacing in Å between charged groups on the polyion
[b] Charge density parameter calculated from Eq. 5 using the dielectric constant of bulk water ($\varepsilon = 78$)
[c] Effective local dielectric constant on the polyion surface
[d] Modified charge density parameter taking into account the local dielectric constant

The value of ε_{eff} for A-100 (**3** with x = 100) is significantly lower than the dielectric constant of bulk water ($\varepsilon = 78$), and it decreases further with decreasing charge density. This behavior is apparently ascribable to the hydrophobicity of the polymer backbone and also to the surface charge (dielectric saturation).

Manning's theory does not take the local effective dielectric constant into consideration, but simply uses the ε value of bulk water for the calculation of ξ. However, since counterion condensation is supposed to take place on the surface of polyions. Manning's ξ should be modified to ξ' by replacing ε with ε_{eff}. The modified parameters ξ' is compared with ξ in Table 1, which leads to the conclusion that the linear charge density parameter calculated with the bulk dielectric constant considerably underestimates the correct one corresponding to the interfacial dielectric constant.

It is clear from Table 1 that pK_{obs} decreases systematically as the charge density of the polyanion decreases, and pK^i shows a slight increase with a decrease in the charge density. The latter arises from a decrease in the effective dielectric constant with decreasing charge density. The electrostatic potential systematically decreases as the charge density decreases.

Figure 4 shows the electrostatic potential calculated as a function of ξ' at distances 5, 10, and 15 Å from the cylinder axis by using the data given in Fig. 1. Experimental points come fairly close to the calculated values at 10 Å, when $\xi' > 1$.

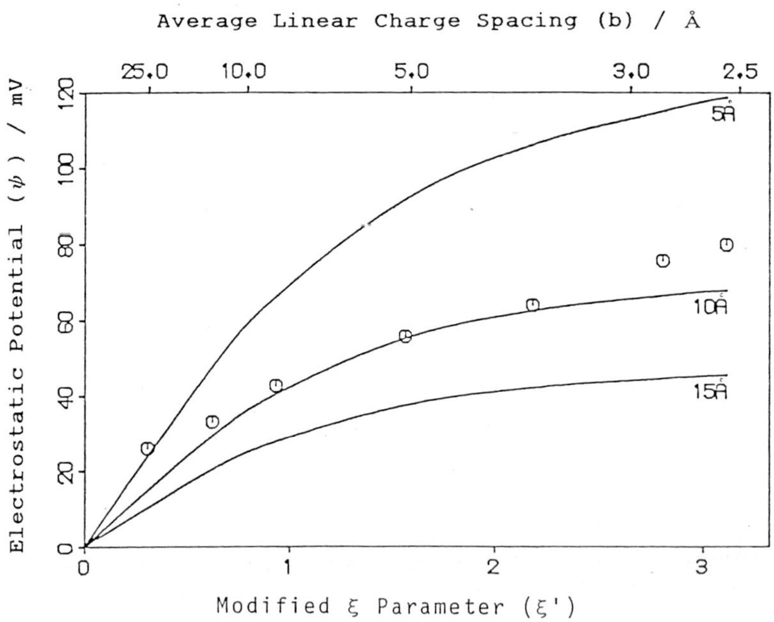

Fig. 4. Relationship between the electrostatic potential calculated from the PB equation and the charge density parameter at distances 5, 10, and 15 Å from the axis of a cylinder with 3 Å radius. Points indicate experimental data listed in Table 1 [42]

This finding may be rationalized for the following reasons. The total length of the Mc pendant moiety from the cylinder axis is approximately 15 Å when the dye moiety is stretched out from the polymer main chain. Since the dye moiety is linked to the main chain via a flexible chemical bond, it may be able to reside at any distance between 3 and 15 Å from the cylinder axis. Thus, on average the Mc residues would experience the polyanion microenvironment at a distance of about 10 A.

Small fractions of a similar type of merocyanine dye moieties (Mc) were also covalently tagged onto poly(sodium 2-acrylamido-2-methylpropanesulfonate) (AMc-3, **4**) and poly[(3-(methacrylamino)propyl)trimethylammonium chloride] (QMc-1, **5**) [49]. The observed pK_{obs} value of 10.92 for AMc-3 was higher than that for the neutral reference (NMc-3, **6**) by 2.24 pH units. By contrast, the

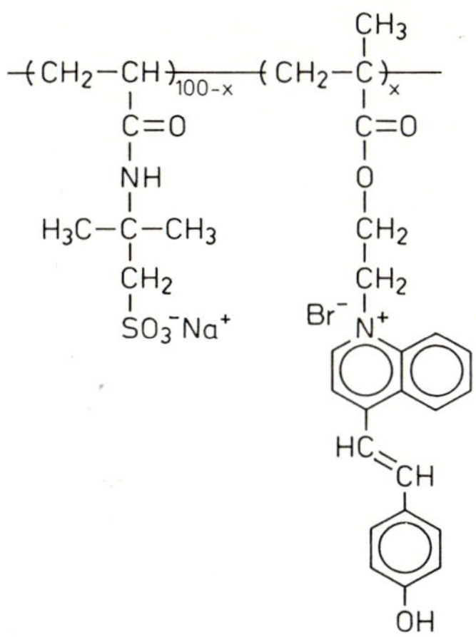

x = 3

AMc-3, **4**

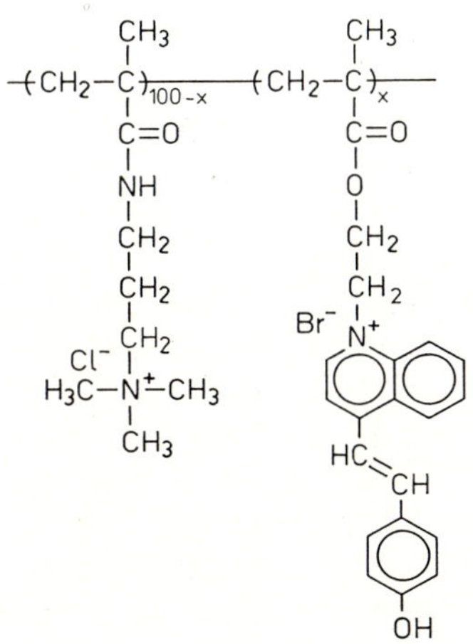

x = 1

QMc-1, 5

corresponding value of 6.18 for QMc-1 was lower than that for NMc-3 by 2.50 pH units. The electrostatic potentials for these polyions were calculated to be −131 and +146 mV for AMc-3 and QMc-1, respectively.

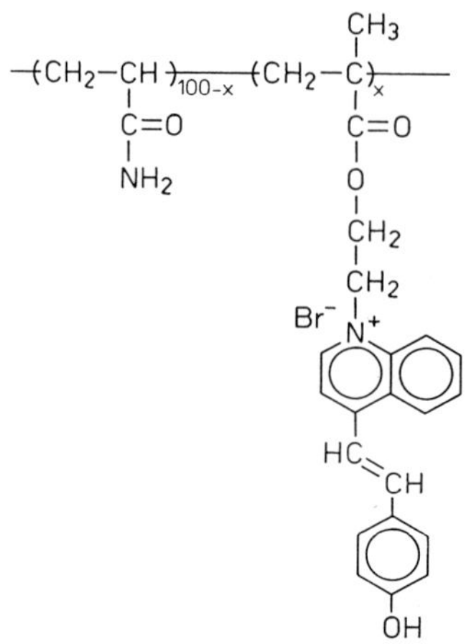

x = 3

NMc-3, **6**

Since the electrostatic potential sharply decreases with increasing distance from the polyelectrolyte cylinder, the degree of reactivity modification by functional groups fixed to the polyion is strongly dependent on the distance from the cylinder surface. Considerable electrostatic potential effects on the photoinduced forward and thermal back electron transfer reactions, which will be discussed in the following chapters, can be attributed to the functional chromophore groups directly attached to the polyelectrolyte back-bone through covalent bonds.

3.3 Behavior of Counterions

According to Manning's theory, Na^+ ions "condense" onto the polyanion until the fraction $(1 - \xi^{-1})$ of the line charge is neutralized when $\xi > 1$; i.e. the effective charge density is lowered to $\xi = 1$ by counterion neutralization [43]. If the carboxylate anions on the polymer chain are electrically neutralized through close contact with Na^+ cations, in other words, if the thickness of the counterion condensation layer on the polyelectrolyte cylinder is less than the distance from the cylinder at which the Mc moieties are situated, then the measured electrostatic potential would become independent of the (original) charge density when $\xi > 1$. This is not the case, but the measured electrostatic potential values progressively increase as the charge density increases. This implies that the condensation layer of Na^+ counter cations is much thicker than the distance between the cylinder surface and the Mc moiety. In other words, most of the counterions appear outside the Mc moieties when viewed from the cylinder surface.

4 Microphase Separation of Amphiphilic Polyelectrolytes

The pioneering work on amphiphilic polyelectrolytes goes back to 1951, when Strauss et al. [25] first synthesized amphiphilic polycations by quaternization of poly(2-vinylpyridine) with *n*-dodecyl bromide. They revealed that the long alkyl side chains attached to partially quaternized poly(vinylpyridine)s tended to aggregate in aqueous solution so that the polymers assumed a compact conformation when the mole fraction of the hydrophobic side chains exceeded a certain critical value. Thus, Strauss et al. became the first to show experimentally the intramolecular micellation of amphiphilic polymers and the existence of a critical content of hydrophobic residues which may be compared to the critical micelle concentration of ordinary surfactants. They called such amphiphilic polyelectrolytes "polysoaps" [25].

Until recent years only a relatively few studies had been reported on the amphiphilic polyelectrolytes. However, several years ago attention began to be directed to the microphase structure as a reaction medium that modifies photophysics and photochemistry [50–64], redox processes [65–67], and chemical reactions [68, 69]. Since then the number of reports on amphiphilic polyelectrolyte systems have increased sharply.

4.1 Solution Behavior of Amphiphilic Polyelectrolytes

Morishima et al. [29–31] prepared amphiphilic copolymers of 2-acrylamido-2-methylpropanesulfonic acid (AMPS) with various hydrophobic comonomers, and studied the tendency of their self-aggregation and the nature of the hydrophobic microdomains thus formed in aqueous solution. Chart 1 shows some of these amphiphilic copolymers. Here, the value of x indicates the mol% content of hydrophobic comonomer units.

The charged segments of the AMPS units in these amphiphilic copolymers effectively solubilize the sequences of hydrophobic monomer units to water. In fact, the copolymers ASt-72 (**7** with x = 72), APh-50 (**8** with x = 50), APy-50 (**9** with x = 50), and ALa-44 (**10** with x = 44) were all soluble in water. The copolymer ACh-x was a little less water soluble: ACh-23 (**11** with x = 23) was almost soluble, whereas ACh-60 was insoluble. All these copolymers were soluble in methanol, *N*, *N*-dimethylformamide, and dimethylsulfoxide, but insoluble in most of other common organic solvents.

The microphase separation of an amphiphilic polyelectrolyte is clearly reflected in the viscosity behavior of its aqueous solution. As a representative example, Fig. 5 shows the reduced viscosities of ASt-x with different styrene (St) content plotted against the polymer concentration in salt-free aqueous solution [29]. The AMPS homopolymer and its copolymers with low St content exhibit negative slopes, which is the typical behavior of polyelectrolytes in the concentration range shown in Fig. 5. With increasing St content, however, the slope systematically decreases and eventually turns to be slightly positive, while reduced viscosity itself markedly decreases. These data indicate that, with increasing St content, the

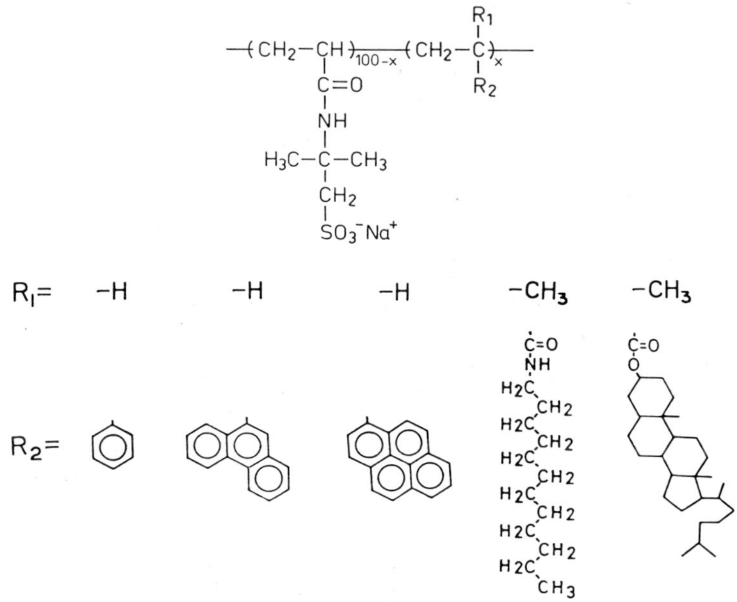

ASt-x, **7** APh-x, **8** APy-x, **9** ALa-x, **10** ACh-x, **11**

Chart 1

polymer coil becomes more compact owing to the hydrophobic self-aggregation of the St residues, which suppresses chain expansion due to segment-segment electrostatic repulsion.

The hydrophobic self-aggregation is more clearly seen in the effect of added salts on reduced viscosity. When KCl is added to an aqueous solution of AMPS homopolymer, reduced viscosity considerably decreases because of the shrinkage of the polymer coil. This salt effect, represented as the ratio of the reduced viscosities in the absence and presence of 0.7 M KCl, is shown as a function of the St content in ASt-x in Fig. 6 [29]. It can be seen that the salt effect on reduced viscosity increases with increasing St content and reaches a maximum at an St content of 55 mol%. This behavior is due to an extra shrinkage of the polymer coil which occurs upon addition of the salt as a result of an increase in hydrophobic interactions between the St groups as well as a decrease in electrostatic repulsions of charged segments with an increase in ionic strength. However, when the St content is increased beyond 55 mol%, reduced viscosity suddenly becomes nearly independent of ionic strength. This behavior indicates that the polymer coil is already compact in salt-free solution owing to an extensive self-aggregation of the St residues.

The formation of a microphase structure leads to a surface-active effect [31]. The surface tension of water is considerably lowered when amphiphilic copolymers are dissolved. The surface-active effect appears more significantly in the copolymers with more hydrophobic units.

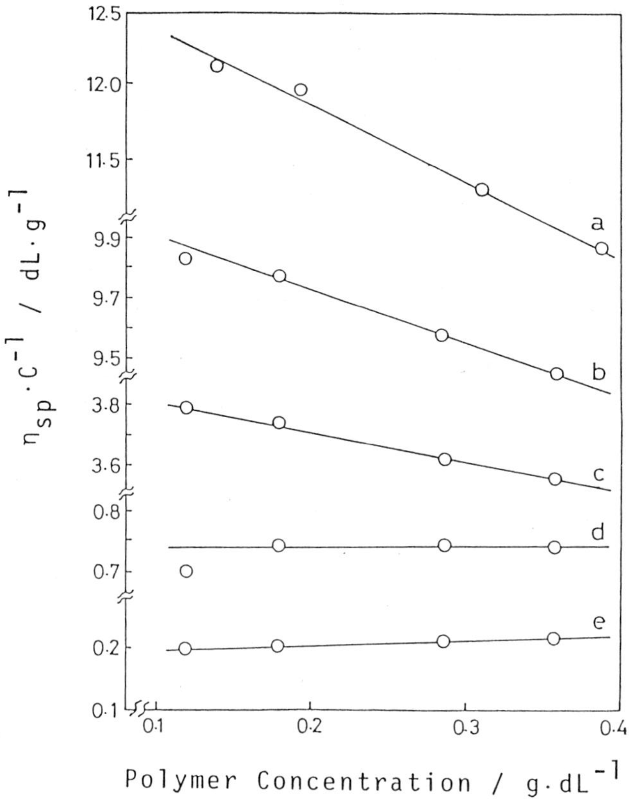

Fig. 5. Plots of reduced viscosity as a function of the concentration of ASt-x (**7**) in pure water at 30 °C; (a) x = 0 (PAMPS), (b) x = 46, (c) x = 54, (d) x = 64, and (e) x = 72 [29]

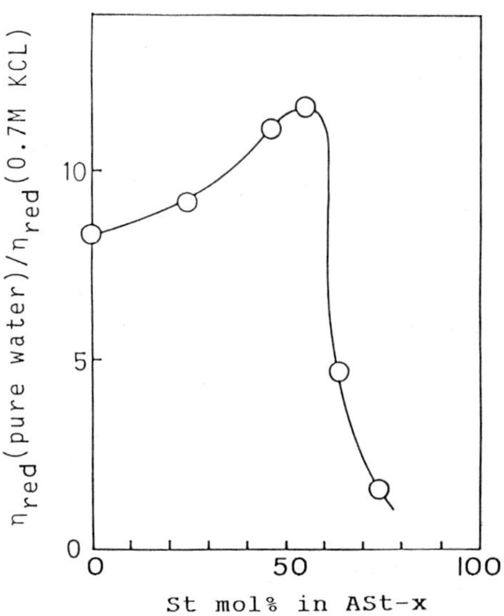

Fig. 6. Effect of added salt on reduced viscosity for ASt-x (**7**) with different St content [29]

The microphase structure was clearly observed in transmission electron micrographs of the film of amphiphilic copolymers cast from aqueous solutions [29, 31]. An important finding was that no microphase structure was observed for the film cast from organic solutions. This difference indicates that a microphase structure is formed in aqueous solution, but not in organic solution. Different hydrophobic groups showed considerably different morphological features; i.e. whether microphase separation leads to a secondary or higher structure depends on the type of hydrophobic units in the copolymers [31].

4.2 Spectroscopic Observations of Microphase Separation

^{1}H NMR spectroscopy is a useful experimental tool for studying hydrophobic self-aggregation. In Fig. 7, ^{1}H NMR peaks for the phenyl protons in ASt-x with different St content are compared [29]. The spectra were measured in DMSO-d_6 and in D_2O at a constant polymer concentration (7 wt%) under the same instrument conditions. In DMSO-d_6, the peak intensity increases systematically with increasing St mol% in the copolymer. In D_2O, by contrast, the peak intensity first increases with an increase in the St content, but begins to decrease as the St content exceeds 55 mol%. Eventually, the peak disappears almost completely at an St content of 72 mol%. This marked line broadening in the D_2O solution can be attributed to a strong restriction of the motion of phenyl rings that occurs when they are incorporated in hydrophobic self-aggregates.

The St content at which the line broadening starts to occur coincides with the one marking the sharp decrease in the salt-effect on reduced viscosity (see Fig. 6).

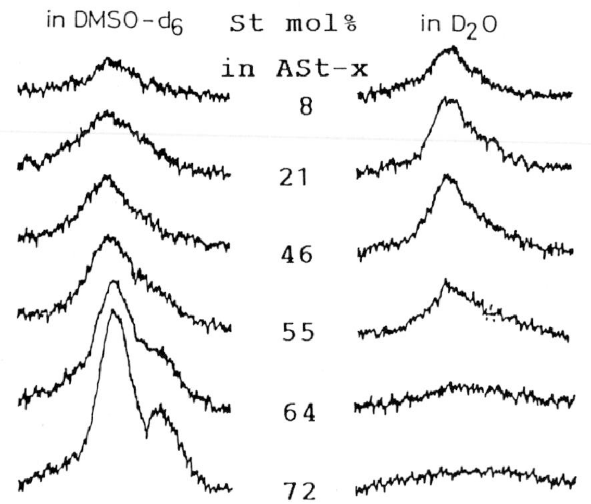

Fig. 7. ^{1}H NMR spectra for ASt-x (7) with different St content in DMSO-d_6 and in D_2O at room temperature. The concentrations of the sample solutions were adjusted to 7 wt% [29]

It represents the critical St content in the copolymer in the sense that it is the onset of progressive self-aggregation of the St groups.

NMR line broadening occurred more pronouncedly for copolymers with aromatic hydrophobic units than for those with aliphatic ones [31]. This implies that polycyclic aromatic rings tend to be more tightly packed when self-aggregating than do aliphatic groups.

The presence of a critical St content in ASt-x can also be seen in fluorescence spectra [29]. This copolymer in aqueous solution shows an excimer emission peaking at 325 nm. As shown in Fig. 8, the intensity of the excimer emission increases, while the monomer emission decreases, with increasing St content. Eventually the excimer dominates the monomer emission at an St content of 72 mol%. The excimer emission becomes apparent at an St content of about 50 mol%, which agrees with the critical St content estimated by viscometry and NMR spectroscopy. The existence of the critical St content suggests the hydrophobic self-aggregation to be a cooperative process.

The formation of a microphase structure can be sensitively detected by using hydrophobic fluorescent probes. Hydrophobic microdomains tend to solubilize hydrophobic small molecules present together in aqueous solution. For example, diphenylhexatriene (DHT) is hydrophobically bound to the St aggregates in ASt-x in aqueous solution and, as a result, the fluorescence intensity is greatly enhanced. Figure 9 shows the fluorescence intensity of DHT in the presence of ASt-x relative to the intensity in its absence (I/I_0) as a function of the ASt-x concentration [29].

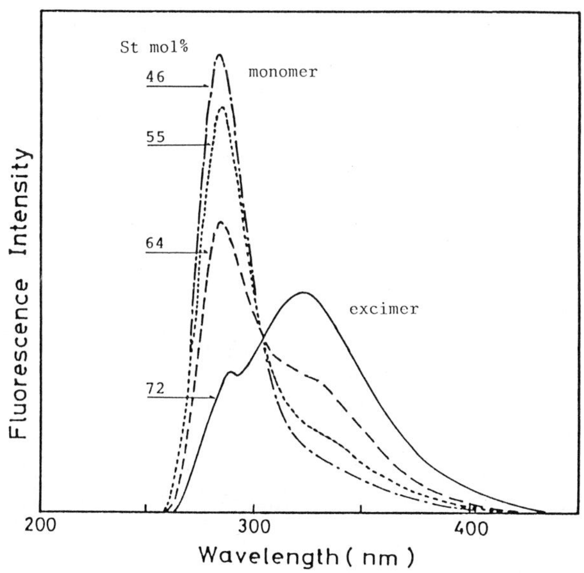

Fig. 8. Fluorescent spectra for ASt-x (**7**) with different St content in pure water; (—·—·—) x = 46, (·····) x = 55, (– – –) x = 64, (———) x = 72; excitation wavelength, 260 nm [29]

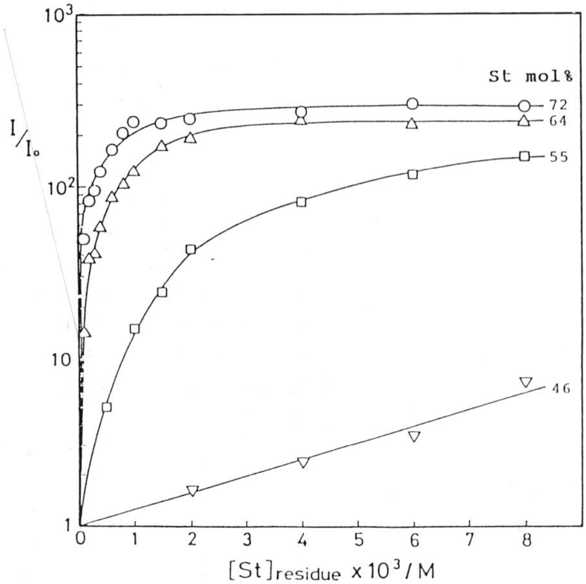

Fig. 9. Enhancement of fluorescence intensity of DHT in aqueous solution in the presence of ASt-x (**7**) with different St content; (▽) x = 46, (□) x = 55, (△) x = 64, (○) x = 72; [DHT] = 8×10^{-7} M; excitation wavelength, 370 nm [29]

A pronounced enhancement of the fluorescence intensity occurs when the St content in ASt-x is higher than 55 mol%, again, indicating that a hydrophobic microdomain is formed at the critical St content.

The theory of hydrophobic interaction [70-72] indicates that hydrophobic residues tend to associate with one another so as to minimize the surface area exposed to the aqueous phase and thereby to release a maximum number of structured water molecules. Therefore, the "steric fit" between the hydrophobic groups may be an important factor for the hydrophobic association. It is reasonable to consider that aromatic hydrophobic groups may undergo tighter hydrophobic self-association because planar aromatic rings would sterically fit with each other to favor the release of structured water.

We will be concerned in Chapter 6 with photoinduced ET of hydrophobic chromophores that are confined to the microphase structure of amphiphilic polyeletrolytes.

5 Electrostatic Effects on Photoinduced Electron Transfer

Meisel et al. [18] and later Rabani et al. [22] investigated photoinduced ET in aqueous solution in the presence of polyeletrolytes. In these early studies, they showed that the rate of ET was kinetically changed and the photoinduced charge separation was assisted to some extent by the addition of a polyeletrolyte. These effects arise from the fact that charged reacting species experience the electrostatic potential produced by the local charges of the added polyeletrolyte. However, chemical species such as photoactive chromophores and electron acceptors (or donors) do not always experience a high electrostatic potential when they are simply mixed with a polyelectrolyte.

In contrast, if photoactive chromophores are covalently tethered to the polyelectrolyte main chain, the chromophore reactivity in the photoinduced ET is much more greatly changed because the reaction sites are totally constrained to the polyelectrolyte molecular surface.

Morishima et al. [30, 50–54, 73–76] have made extensive investigations on photoinduced ET, using polycyclic aromatic chromophores covalently attached to polyelectrolytes. They were the first to show that the polyeletrolyte molecular surface provides an unusual microenvironment which greatly changes the rate of photoinduced ET and the fate of the charged photoproducts.

5.1 Forward Electron Transfer as Studied by Fluorescence Quenching

The fluorescence from the phenanthrene (Phen) residues covalently tethered to a polyanion (APh-x, **8**) or a polycation (QPh-x, **12**) was quenched by methylviologen (MV^{2+}, **13**) and also by a zwitterionic viologen (SPV, **14**) (4,4′-bipyridinium-1,1′-bis(trimethylenesulfonate) in aqueous solution [74]. In these cases, the quenching occurs via ET from the singlet-excited phenanthrene to the viologens, and thus it provides a practical means for studying the rate of the photoinduced forward ET from the singlet-excited phenanthrene.

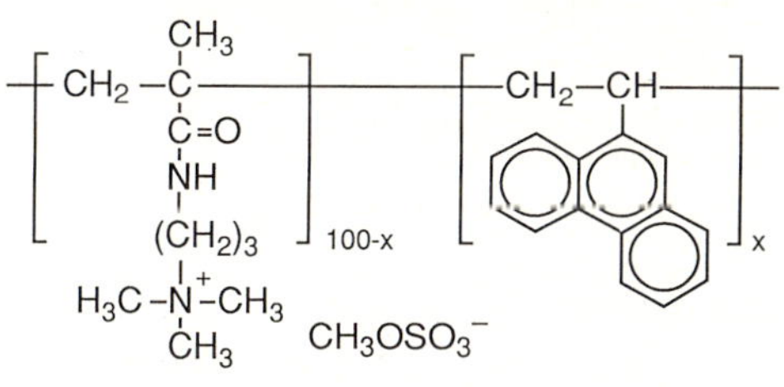

QPh-x, **12**

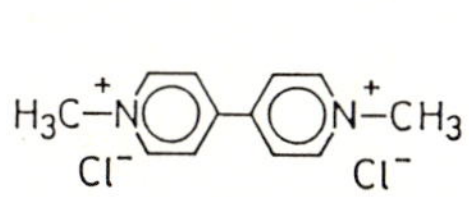

MV^{2+}, **13**

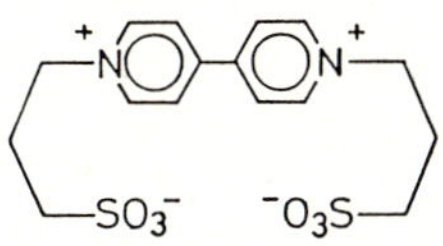

SPV, **14**

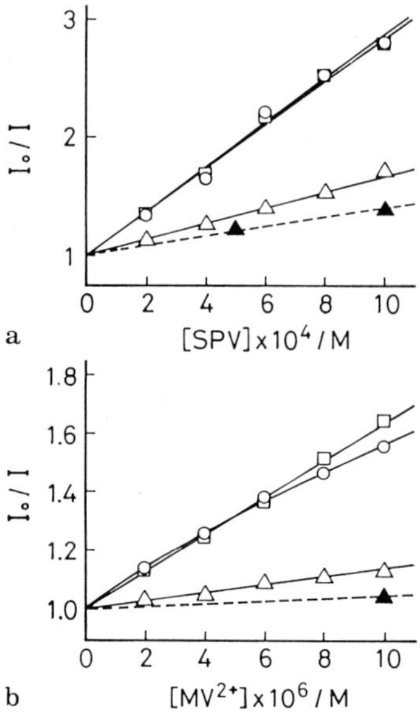

Fig. 10 a, b. Stern-Volmer plots for APh-x (**8**) with (**a**) SPV (**14**) and (**b**) MV^{2+} (**13**) in aqueous solution; (○) APh-50, (□) APh-32, (△) APh-9, (▲) APh-9 in aqueous NaCl (0.2 M) solution [74]

Figure 10 illustrates Stern-Volmer plots for the fluorescence quenching of APh-x by MV^{2+} and SPV in aqueous solution [74]. With MV^{2+}, the quenching is so effective that it occurs at very low quencher concentrations (in the range of 10^{-6} M), whereas with SPV, it proceeds to about the same extent at two-orders of magnitude higher quencher concentration (in the range of 10^{-4} M).

Table 2 lists the apparent Stern-Volmer quenching constants (K_{SV}) for APh-x, estimated from the initial slopes of the Stern-Volmer plots, along with the

Table 2. Rate constants for the fluorescence quenching of APh-8 (**8**) by SPV (**14**) and MV^{2+} (**13**) in aqueous solution [74]

Sample[a]	Additives	τ/ns[b]	K_{SV}[c]/M^{-1}		$k_q \times 10^{-9}/M^{-1}\,s^{-1}$	
			SPV	MV^{2+}	SPV	MV^{2+}
APh-50		33	1850	69000	56	2100
APh-50	PAMPS (2 mM)	33	925	13900	28	420
APh-32		38	1810	36900	48	1700
APh-9		38	675	14300	17	380
APh-9	NaCl (0.2 M)	38	351	5550	9.2	146
Phenanthrene/SDS[d]		40	710	1600	ca. 20	ca. 40
AM		34	472	1070	14	31
AM	NaCl (0.2 M)	34	458	530	13	16

[a] [Phen] (residue) = 0.2 mM.
[b] Average lifetime for fluorescence from the Phen residue.
[c] Calculated from the initial slope of the Stern-Volmer plot.
[d] Phenanthrene solubilitzed in the SDS micelle; [SDS] = 50 mM

second-order equivalent quenching rate constants (k_q). For comparison, the data for an anionic surfactant micellar system (phenanthrene solubilized in sodium dodecyl sulfate (SDS) micelles) and for a monomer model (AM, **15**) are also presented in Table 2. The k_q values for MV^{2+} are of the order of $10^{12}\ M^{-1}\ s^{-1}$, which far exceed the diffusion-controlled rate constant (ca. $7 \times 10^9\ M^{-1}\ s^{-1}$) [74].

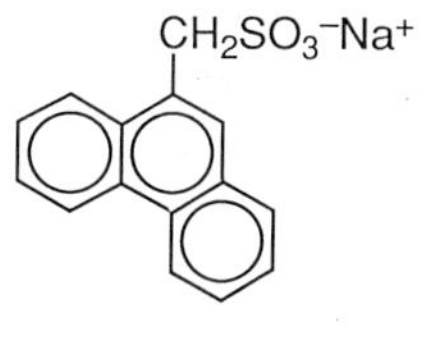

AM, **15**

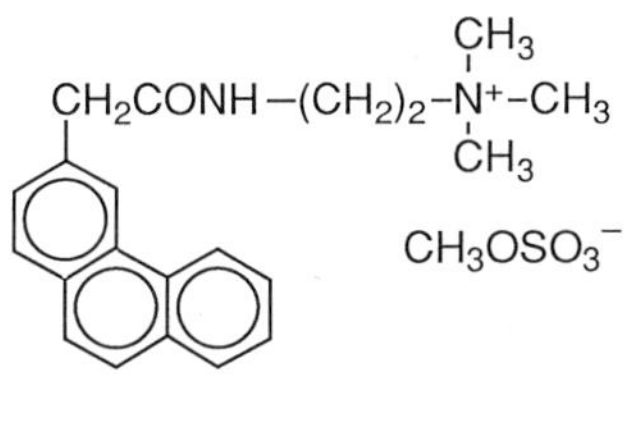

QM, **16**

These findings can be attributed to the increase in the local concentration of MV^{2+} on the APh-x molecular surface caused by eletrostatic interactions. In contrast, the quenching constants for MV^{2+} and SPV show no such large difference in the SDS micellar and AM systems. The addition of NaCl reduces the value of k_q to about one-third that for the quenching of APh-9 (APh-x with 9 mol% Phen units) by MV^{2+} in a salt-free solution. This effect is mainly accounted for by the screening of electrostatic attraction between APh-9 and MV^{2+}.

On the other hand, the k_q values for the quenching of APh-x by SPV are of the order of $10^{10}\ M^{-1}\ s^{-1}$; two orders of magnitude smaller than those for MV^{2+}. The k_q value for SPV appears to be somewhat larger than the diffusion-controlled limit, suggesting that the zwitterionic SPV molecules are concentrated on the surface of APh-x in such a way that the positive charges on SPV come nearer the APh-x anionic segments. This suggestion is consistent with a decrease in the k_q value for SPV by addition of NaCl. For example, the additions of 0.2 M NaCl decreased k_q for APh-9 from 1.7×10^{10} to $2 \times 10^9\ M^{-1}\ s^{-1}$, as shown in Table 2.

The fluorescence quenching depends on the content of the Phen units (the x values) in APh-x. An aqueous solution of APh-9 contained as many charged groups (SO_3^-) as about 10 times that of APh-50, when compared at the same molar concentration of the Phen residues. When AMPS homopolymer (PAMPS) was added to a solution of APh-50 so that the SO_3^- residue concentration was equal to that for APh-9, the k_q value for the APh-50 quenching by MV^{2+} decreased from 2.1×10^{12} to $4.2 \times 10^{11}\ M^{-1}\ s^{-1}$, which is close to the k_q value for APh-9 (Table 2). From these facts the lower k_q values for APh-x with lower x (higher

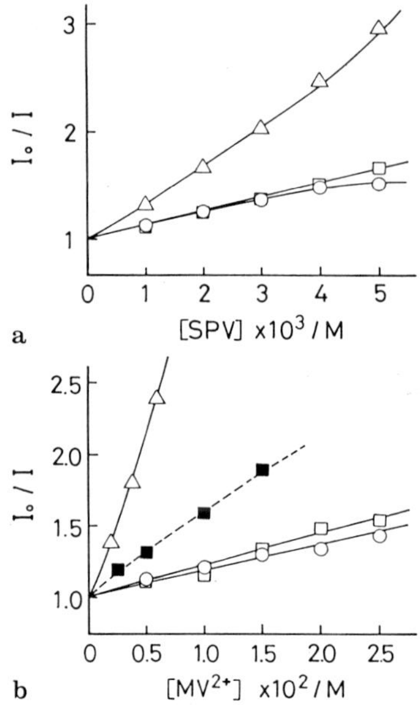

Fig. 11a, b. Stern-Volmer plots for QPh-x (**12**) and QM (**16**) with (**a**) SPV (**14**) and (**b**) MV^{2+} (**13**) in aqueous solution: (○) QPh-47, (□) QPh-14, (△) QM, (■) QPh-14 in aqueous NaCl (0.2 M) solution [74]

mole fractions of the SO_3^- unit) may be associated with a decrease in the effective local concentration of MV^{2+} around the Phen site on APh-x. This decrease arises from the electrostatic association of MV^{2+} with the SO_3^- segments remote from the Phen residues. Importantly, fluorescence quenching is more effective for APh-x than for phenanthrene solubilized in SDS micelles.

Figure 11 shows Stern-Volmer plots for fluorescence quenching of the amphiphilic cationic copolymer QPh-x [74]. The quenching of QPh-x with MV^{2+} is expected to be much less effective than that of APh-x. The quenching data for the QPh-x system are presented in Table 3. For comparison, the data for a related

Table 3. Rate constants for the fluorescence quenching of QPh-x (**12**) by SPV (**14**) and MV^{2+} (**13**) in aqueous solution [74]

Sample[a]	Additives	τ/ns[b]	K_{SV}[c]/M^{-1}		$k_q \times 10^{-9}/M^{-1}\,s^{-1}$	
			SPV	MV^{2+}	SPV	MV^{2+}
QPh-47		34	130	19	3.8	0.56
QPh-14		42	130	22.4	3.1	0.53
QPh-14	NaCl (0.2 M)	42	127	74	3.0	1.7
Phenanthrene/CTAB[d]		17	21.8	0	ca. 1	0
QM		44	320	190	7.3	4.3
QM	NaCl (0.2 M)	44	370	320	8.4	7.3

[a] [Phen] (residue) = 0.2 mM.
[b] Average lifetime for fluorescence from the Phen residue.
[c] Calculated from the initial slope of the Stern-Volmer plot.
[d] Phenanthrene solubilized in the CTAB micelle; [CTAB] = 5 mM

monomer model (QM, **16**) and for a cationic micelle system are also given in Table 3. With MV^{2+}, the k_q value for QM is close to the diffusion limit, whereas the k_q values for QPh-x are significantly smaller than that. This difference can be attributed to the difficulty for MV^{2+} in reaching the polyeletrolyte surface owing to the positively charged QPh-x segments, as evidenced by the salt-effect on fluorescence quenching shown in Table 3. It is to be noted that the fluorescence of phenanthrene solubilized in CTAB micelles is not quenched by MV^{2+}. Thus, it appears that the surface charges of the CTAB micelle more effectively prevent the access of MV^{2+} to the chromphore than those of QPh-x.

More recently, several groups have investigated electrostatic effects on the fluorescence quenching of hydrophobic chromophores covalently attached to various polyanions. The photophysics of the chromophores incorporated in the polyeletrolytes at small mole fractions is relatively simple, because no interaction is expected to occur between the incorporated chromophores. For this reason, most of the studies have focused on amphiphilic polyeletrolytes loaded with a low amount of hydrophobic chromophores.

Thomas et al. [64] prepared copolymers of 1-pyreneacrylic acid with acrylic acid (AA), **17**, and with methacrylic acid (MA), **18**, incorporating not more than one pyrene unit per polymer chain. The mole ratios of pyrene to AA and MA units were 1150 and 1390, respectively. The molecular weights of **17** and **18** were 8.3×10^4

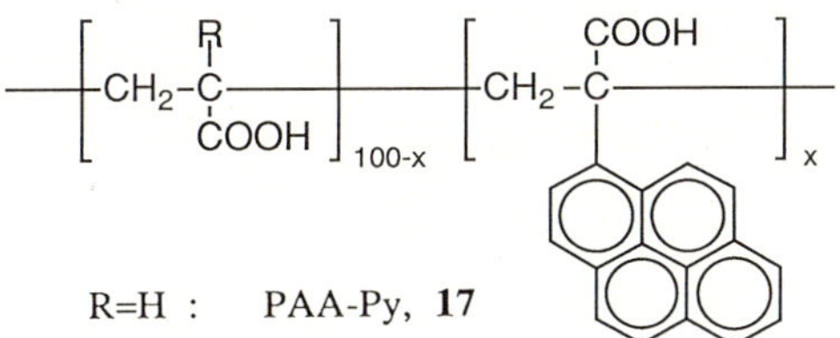

R=H : PAA-Py, **17**

R=CH_3 : PMA-Py, **18**

and 1.2×10^5, respectively. Positively charged fluorescence quenchers such as Tl^+ and Cu^{2+} ions exhibited a very high quenching efficiency when the AA and MA units were ionized at basic pH's. For example, the k_q value for the quenching of pyrene fluorescence with Cu^{2+} was 4.5×10^{10} M^{-1} s^{-1} for **17** at pH 7.2. With a negatively charged quencher, by contrast, the quenching efficiency was very low; e.g., the $k_q$value with I^- anions was 5.2×10^7 M^{-1} s^{-1} for **18** at pH 11.8.

Similar data were reported by Turro et al., [62, 63] who synthesized a copolymer of AA with 1.5 mol% of 2-[4-(1-pyrene)butanoyl]aminopropenoic acid, **19** and studied the fluorescence quenching with Tl^+, Cu^{2+}, and I^- ions in aqueous solution.

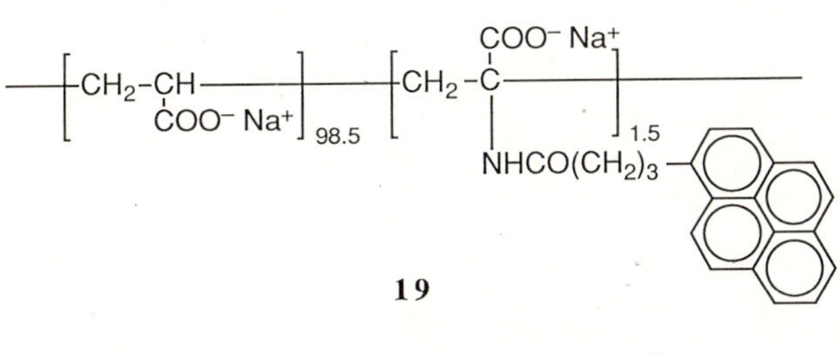

19

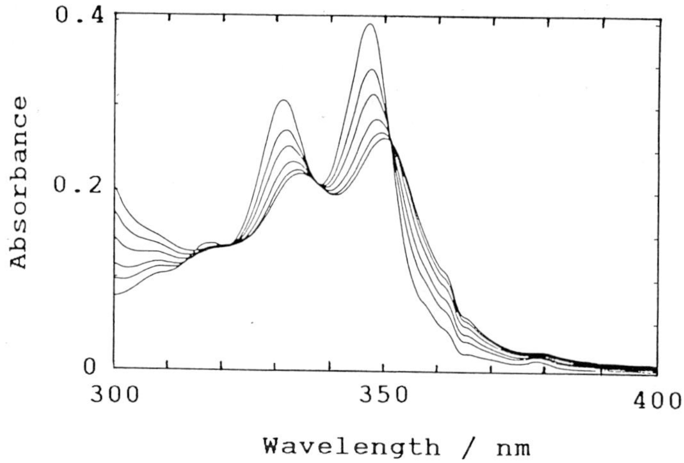

Fig. 12. Perturbed absorption spectra for PMAvPy (**20**) in the presence of MV^{2+} (**13**). Concentrations of MV^{2+} from top to bottom; $0.5 \times 10^{-6}, 1 \times 10^{-5}, 3 \times 10^{-5}, 4 \times 10^{-5}$ M [77]

Webber et al. [77] have extensively studied photoinduced ET of hydrophobic chromophores covalently tethered to various polyelectrolytes. They prepared copolymers of MA, AA, and styrenesulfonate containing small mole fractions of 1-vinylpyrene. The content of the pyrene units was 0.624, 0.319, and 0.412 mol% for PMAvPy (**20**), PAAvPy (**21**), and PSSvPy (**22**), respectively. Consistent with the observations described above, the quenching of excited pyrene moiety by cationic quenchers was extremely efficient. An important feature of the fluorescence quenching by MV^{2+} is that MV^{2+} forms a nonfluorescent CT complex with the pyrene residue, leading to highly efficient static quenching. Figure 12 represents

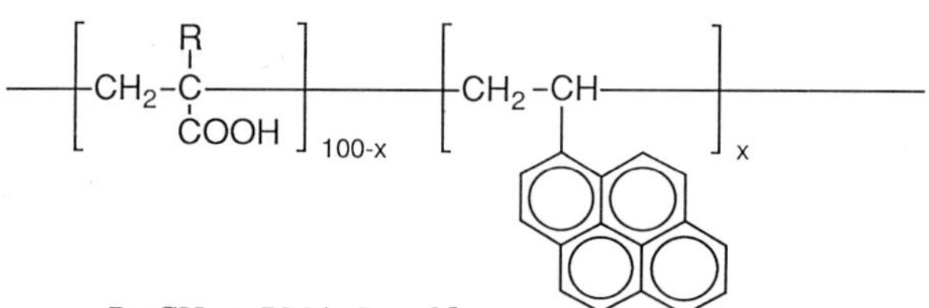

R=CH_3 : PMAvPy, **20**

R=H : PAAvPy, **21**

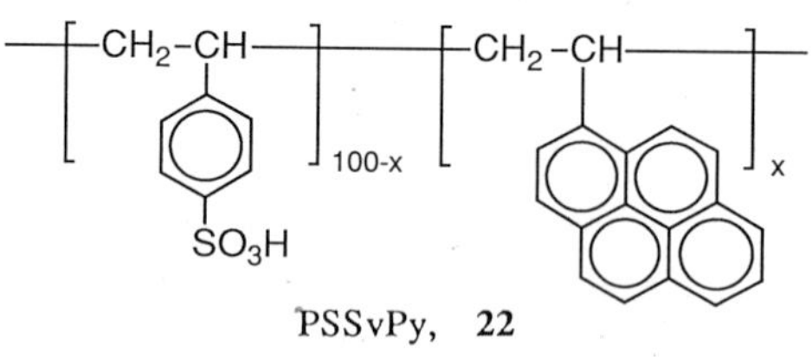

PSSvPy, **22**

an example of the changes in the absorption spectra of PMAvPy due to the CT complexation upon addition of MV^{2+}. By using the modified Benesi-Hildebrand equation the CT formation constant (K_{CT}) can be estimated (Table 4). The CT complexation is greatly facilitated with MV^{2+} as compared to SPV, owing to the electrostatic attraction between the cation and the polyanion.

If it is assumed that the ground-state complex is nonfluorescent and the complexation follows a single equilibrium

$$C + Q \overset{K_0}{\rightleftharpoons} CQ \tag{7}$$

the following relations can be obtained for the case $[Q] = [Q]_0$:

$$[CQ]/[C]_0 = K_0[Q]_0/(1 + K_0[Q]_0) \cong K_0[Q]_0 \tag{8}$$

where C represents a chromophore, Q a quencher, $[C]_0$ the concentration of the chromophore before complexation, and $[Q]_0$ the initial concentration of the quencher. The last relation in eq 8 is valid for $K_0[Q]_0 \ll 1$. The chromophore C is assumed to be quenched by the usual diffusive Stern-Volmer mechanism. The I/I_0 ratio is then given by

$$\begin{aligned} I/I_0 &= ([C]_0 - [CQ])/(1 + K_{SV}[Q])\,[C]_0 \\ &= (1 - K_0[Q]_0)/(1 + K_{SV}[Q]_0) \end{aligned} \tag{9}$$

where $1 + K_{SV}[Q]$ represents the dynamic quenching.

The values of K_0 and K_{SV} obtained by using eq 9 are given in Table 4. The ground-state complex considered in eq 7 includes not only the CT complex but all kinds of complexes that may lead to apparent static quenching. Therefore, usually K_0 is larger than K_{CT} as can be seen from Table 4 (although there are a few exceptions).

Webber et al. [60, 78] also studied the fluorescence quenching of diphenylanthracene (DPA) covalently bound to poly(methacrylic acid), PMAvDPA (**23**) [60], and to sodium poly(styrenesulfonate), PSSvDPA (**24**)[78]. The fluorescence quenching of the excited DPA moiety by MV^{2+} and Cu^{2+} was also highly efficient. For example, with PMAvDPA of 0.073 mol% DPA content, the k_q values at pH

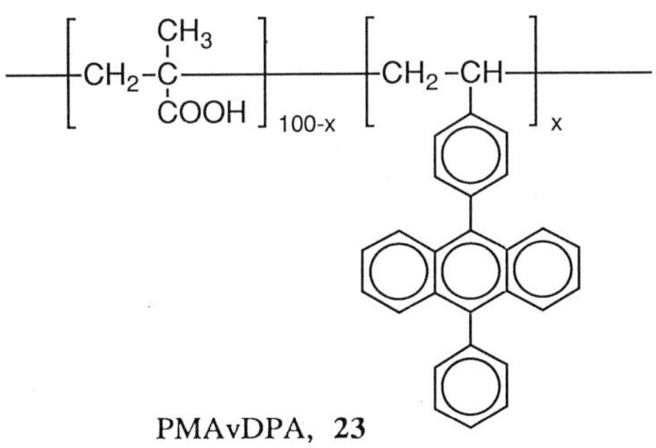

PMAvDPA, **23**

Table 4. K_{CT}, K_{SV} and K_0 values for the polyelectrolyte-bound pyrene system [77]

Polymer	pH	Quencher	K_{CT}/M^{-1}	K_0/M^{-1}	K_{SV}/M^{-1}
PMAvPy (**20**)	11	MV^{2+}	7.8×10^4	3.4×10^4	1.2×10^5
PMAvPy (**20**)	11	SPV	6.3×10^2	8.0×10^1	8.4×10^2
PAAvPy (**21**)	10	MV^{2+}	3.5×10^4	5.8×10^3	6.9×10^4
PAAvPy (**21**)	10	SPV	8.1×10^2	8.1×10^2	1.2×10^3
PSSvPy (**22**)		MV^{2+}	7.6×10^3	2.3×10^3	2.0×10^4
PSSvPy (**22**)		SPV	5.6×10^2	2.1×10^2	1.9×10^3

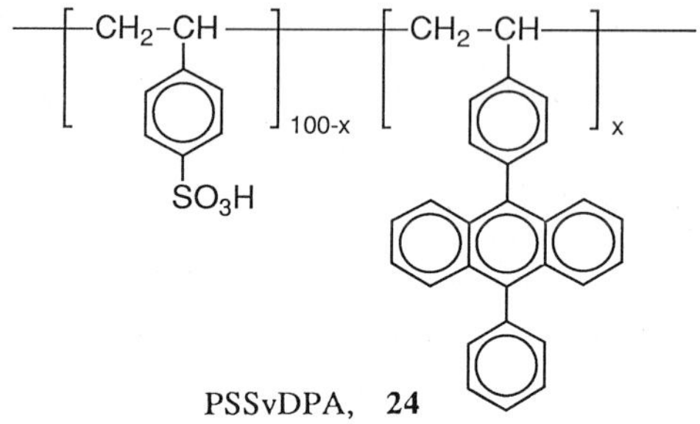

PSSvDPA, **24**

9.0 for MV^{2+} and Cu^{2+} were 6.6×10^{10} and 5.8×10^{10} M^{-1} s^{-1}, respectively [60]. For PSSvDPA with 2.6 mol% DPA content, the quenching efficiency was extremely high; e.g., the k_q values for MV^{2+} and Cu^{2+} were 1.4×10^{13} and 6×10^{12} M^{-1} s^{-1}, respectively [78]. To account for efficient quenching by MV^{2+}, Webber et al. [60, 78] have postulated a kinetic model in which the preferential association of MV^{2+} is assumed to occur as a result of a combined effect of electrostatic, hydrophobic, and CT interactions of the quencher with the DPA moieties. The large difference in k_q between PMAvDPA and PSSvDPA seems to be ascribable to the difference in the DPA content between the copolymers, as discussed earlier in this chapter.

The DPA moiety is less active in forming the CT complex with viologens than the pyrene moiety; e.g., for PMAvDPA the K_{CT} values with MV^{2+} and SPV are 1.3×10^3 M^{-1} and almost zero, respectively, at pH 8–9 [60, 77], whereas for PMAvPY they are 7.8×10^4 and 6.3×10^2 M^{-1}, respectively, at pH 11 [77]. Therefore, the polymer-bound pyrene system undergoes much more static quenching than the polymer-bound DPA system. As will be discussed in Chapter 6, it is very important for charge separation whether the fluorescence quenching is static or dynamic.

Tris(2,2′-bipyridine)ruthenium(II) complex ($Ru(bpy)_3^{2+}$) has been most commonly employed as a chromophore in the studies of photoinduced ET. Electrostatic effects on the quenching of the emission from the Ru(II) complex covalently bound to polyeletrolytes have been studied by several groups [79–82].

Kaneko et al. [80, 81] prepared copolymers of AA (93.9–95.9 mol%) and vinylbipyridine (1.6–3.7 mol%) with pendant $Ru(bpy)_3^{2+}$ (2.4–2.5 mol%) (**25**). The quenching of the excited state of the pendant Ru(II) complex by MV^{2+} was accelerated in alkaline aqueous solution owing to the electrostatic attraction of the cationic quencher. Interestingly, the quenching efficiency was dependent on the molecular weight of **25**. The quenching of the polymer with MW 2100 occurred

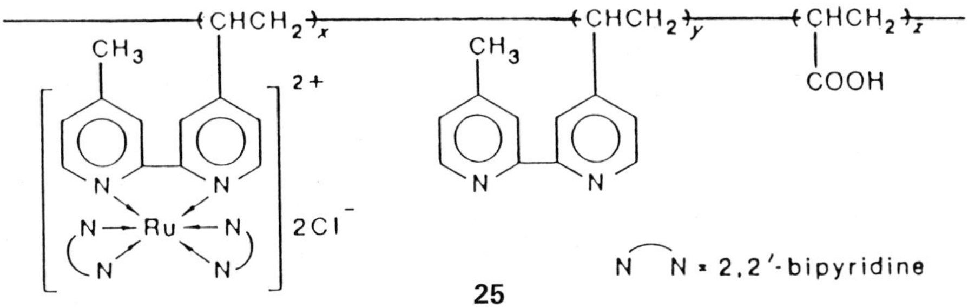

25

via a dynamic mechanism, whereas that of higher molecular weight polymers (MW 4400 and 13300) occurred via a mixed dynamic and static process with a much higher quenching efficiency [81]. Kaneko et al. proposed that these findings could be explained with kinetic models which assumed multistep association between the quencher and the anionic domains formed around the pendant Ru(II) complex [81].

5.2 Back Electron Transfer as Studied by Laser Photolysis

Morishima et al. [75, 76] have shown a remarkable effect of the polyelectrolyte surface potential on photoinduced ET in the laser photolysis of APh-x (**8**) and QPh-x (**12**) with viologens as electron acceptors. Decay profiles for the SPV (**14**) radical anion ($SPV^{-\cdot}$) generated by the photoinduced ET following a 347.1-nm laser excitation were monitored at 602 nm (Fig. 13) [75]. For APh-9, the $SPV^{-\cdot}$ transient absorption persisted for several hundred microseconds after the laser pulse. The second-order rate constant (k_b) for the back ET from $SPV^{-\cdot}$ to the oxidized Phen residue ($Phen^{+\cdot}$) was estimated to be 8.7×10^7 M^{-1} s^{-1} for the APh-9-SPV system. For the monomer model system (AM(**15**)-SPV), on the other hand, k_b was 2.8×10^9 M^{-1} s^{-1}. This marked retardation of the back ET in the APh-9-SPV system is attributed to the electrostatic repulsion of $SPV^{-\cdot}$ by the electric field on the molecular surface of APh-9. The addition of NaCl decreases the electrostatic interaction. In fact, it increased the back ET rate. For example, at NaCl concentrations of 0.025 and 0.2 M, the value of k_b increased to 2.5×10^8 and 5.2×10^8 M^{-1} s^{-1}, respectively.

Retardation of back ET was also observed with phenanthrene solubilized in the SDS micelle ($k_b = 6.8 \times 10^7$ M^{-1} s^{-1}) (see Fig. 13) [75]. However, as can be seen from Fig. 13, the transient yield of $SPV^{-\cdot}$ for the micellar system is extremely low, presumably because only a small fraction of $SPV^{-\cdot}$ can escape from the geminate ion pair. This finding implies that SPV preferably resides inside the micelle and that the electron transfer mainly takes place in the micelle, not across the charged surface.

For QPh-x, the decay of $SPV^{-\cdot}$ was faster than that for the APh-x-SPV system, and obeyed first-order kinetics [76]. The addition of NaCl (0.2 M) caused the decay profile to change; i.e. the back ET was considerably slowed and the decay kinetics changed from first-order to second-order, with a reaction rate constant of $k_b = 1.8 \times 10^9$ M^{-1} s^{-1}. These findings suggest the impossibility of escape of

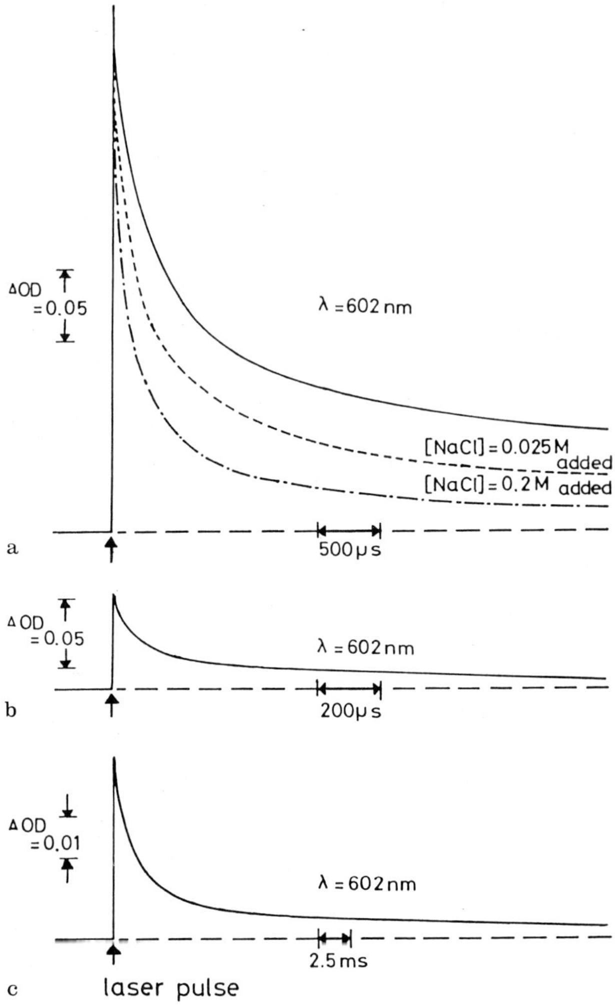

Fig. 13a–c. Absorption decays of $SPV^{-\cdot}$ monitored at 602 nm after laser excitation of (**a**) APh-9 (**8**), (**b**) AM (**15**), and (**c**) phenanthrene solubilized in the SDS micelle; [Phen] = 2.5 mM, [SPV] = 10 mM [75]

$SPV^{-\cdot}$ from the electric field of the polycation, which leads to a first-order back ET kinetics. Since the addition of NaCl interferes with the electrostatic binding of $SPV^{-\cdot}$ by QPh-14, $SPV^{-\cdot}$ can escape into the bulk phase by diffusion. Therefore, the back ET occurs via a bimolecular process when NaCl is added.

For the QPh-x-MV^{2+} system, the methylviologen cation radical ($MV^{+\cdot}$) generated by laser photolysis decayed with a rate constant of $k_b = 3.2 \times 10^8\ M^{-1}\ s^{-1}$. This relatively strong retardation of the back ET is due to the electrostatic repulsion of $MV^{+\cdot}$ by the polycation [76].

In the APh-x-MV^{2+} system, the electrostatic potential of APh-9 suppresses the escape of $MV^{+\cdot}$ from the polyanion, leading to a first-order back ET kinetics.

Here, the addition of NaCl also changes the decay kinetics from first-order to second-order.

In the APh-x-MV^{2+} system, both the growth rate of $MV^{+\cdot}$ in the nanosecond laser photolysis and its fluorescence quenching rate were very fast (details will be discussed in the following section). In the QPh-x-MV^{2+} system, however, the growth of $MV^{+\cdot}$ after the laser pulse consisted of two distinctive components. An initial fast rise was followed by a slow buildup in the microsecond time scale. The initial stage completed in the duration time of nanosecond laser pulse (ca. 20 ns). The subsequent stage reflects the ET from the triplet-excited state of the Phen residue (^{3}Phen*) to MV^{2+}, which is energetically possible [76].

Another distinctive feature of the QPh-x-MV^{2+} system, as compared to the APh-x-MV^{2+} system, is the transient yield which is considerably high, as shown in Table 5 [76]. The maximum transient absorbance at 602 nm due to $MV^{+\cdot}$ is 0.78 for the QPh-14-MV^{2+} system, while it is <0.2 for the APh-9-MV^{2+} system. A point to be noted here is the fluorescence quenching in the QPh-x system being far less effective than that in the APh-x system (the k_q values are 5.3×10^8 M^{-1} s^{-1} and 3.8×10^{11} for the QPh-14-MV^{2+} and APh-9-MV^{2+} systems, respectively).

These facts show that the effectiveness of fluorescence quenching does not change parallel with the yield of transient photoproducts (Table 5). This behaviour can be explained in terms of a very adverse effect of ground-state CT interactions between Phen and MV^{2+} on the charge separation of the photoproducts. In the APh-x-MV^{2+} systems, the CT complexation is facilitated owing to the electrostatic attraction of MV^{2+} with APh-x. In contrast, repulsive positive charges in QPh-x prevent MV^{2+} from approaching the Phen residues at a distance close enough to form the CT complex. For example, for a solution containing a 2.5 mM concentration of the Phen residue and a 1 mM concentration of MV^{2+}, the absorbance at 390 nm due to the Phen-MV^{2+} CT complex is 0.86 and zero for the APh-9-MV^{2+} and QPh-14-MV^{2+} systems, respectively (Table 5) [76]. We may

Table 5. Effect of ground-state CT complexation on fluorescence quenching and the transient yield of $MV^{+\cdot}$ for APh-x (**8**), QPh-x (**12**), and their monomer models AM (**15**) and QM (**16**) in aqueous solution [76]

Sample[a]	Additives	$A_{390\,nm}$[b] (CT band)	k_q[c] $\times 10^{-9}$ M^{-1} s^{-1}	$A_{602\,nm}$[d] (transient)
APh-9		0.86	380	<0.02
APh-9	NaCl (0.2 M)	0.63	146	0.12
AM		1.59	31	0.02
Phenanthrene/SDS[e]		1.06	40	<0.01
QPh-14		0	0.53	0.78
QPh-14	NaCl (0.2 M)	<0.05	1.7	0.45
QM		0.36	7.3	0.38

[a] [Phen] (residue) = 2.5 mM, [MV^{2+}] = 1 mM.
[b] Absorbance of the CT band at 390 nm.
[c] Second-order equivalent rate constant for fluorescence quenching.
[d] Maximum transient absorbance for $MV^{+\cdot}$ at 602 nm after laser pulse.
[e] Phenanthrene solubilized in ther SDS micelle; [SDS] = 50 nM

thus conclude that the ground-state complexation much disfavors the yield of photoproducts, although it highly favors the fluorescence quenching which follows a static mechanism.

Scheme 1 represents the kinetics of a photoinduced ET system including ground-state complexation. Within the DA complex an almost simultaneous back-reaction would occur (step 1). Therefore, the CT complexation causes the yield of the photoproducts to decrease. In this scheme, ($D_S^+ \ldots A_S^-$) denotes a

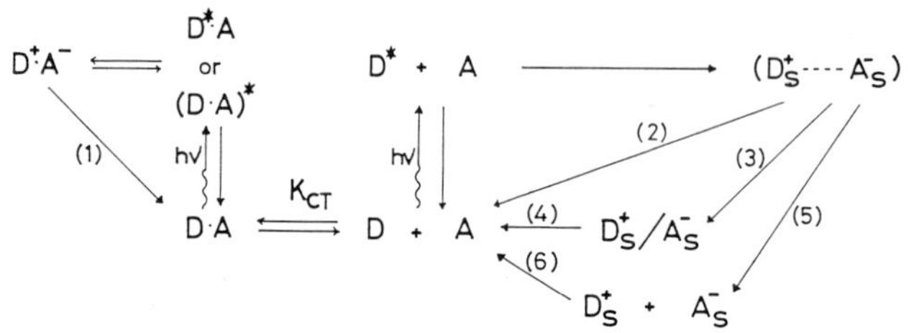

Scheme 1

geminate ion pair which, in usual cases, undergoes either charge recombination (step 2) or dissociation into free ions D_S^+ and A_S^- (step 5). In the absence of subsequent redox reactions, a bimolecular back ET (step 6) will be the fate of these free ions.

In the APh-x-MV^{2+} and QPh-x-SPV systems, the ion pair cannot dissociate into free ions, but remains in the bound state, D_S^+/A_S^-, because A_S^- cannot escape from the interfacial electric field of the polyelectrolyte. In this situation the kinetics of the back ET tends to be first order (step 4) [76]. The ion pair may be either stabilized or destabilized by the interfacial eletrostatic potential, depending on the sign of the electric charges. In the APh-x-SPV system, step 5 may be facilitated so that free ions are stabilized against the back-reaction (step 6). It is important to note here that, even though the photoinduced ET reaction sites are fixed to the polyeletrolyte molecular surface, the very fast back ET within the photochemical cage is not affected by electrostatic potential. In general, the overall quantum yield of the photoinduced ET is largely dependent on this primary process. The transient quantum yields for the $MV^{+\cdot}$ production in the APh-9-SPV and QPh-14-Mv^{2+} systems were approximately estimated to be 0.06 and 0.1, respectively [76]. These low yields are virtually due to the fast charge recombination in the primary ion-pair state. We will discuss this point in more detail in the following chapter.

Sassoon and Rabani [79, 83] constructed an intriguing photoinduced ET system in which the back ET was greatly retarded by the electrostatic repulsion between two different polycations. They prepared poly(3,3-ionene) covalently linked with $Ru(bpy)_3^{2+}$ (**26**) and with an *N,N,N′,N′*-tetraalkyl-*p*-phenylenediamine derivative (**27**). The latter is an electron donor quencher toward the photoexcited Ru(II) complex.

In an aqueous solution containing **26** and **27** the excited state of the Ru(II) complex in **26** essentially has no chance to be directly quenched by the donor quencher in **27**, because a strong electrostatic repulsion acts between **26** and **27**. Sassoon and Rabani added methoxydimethylaniline (MDMA, **28**) to this system

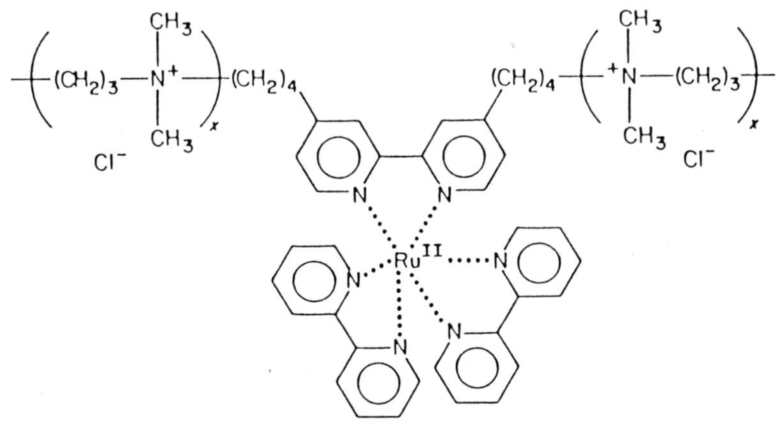

26

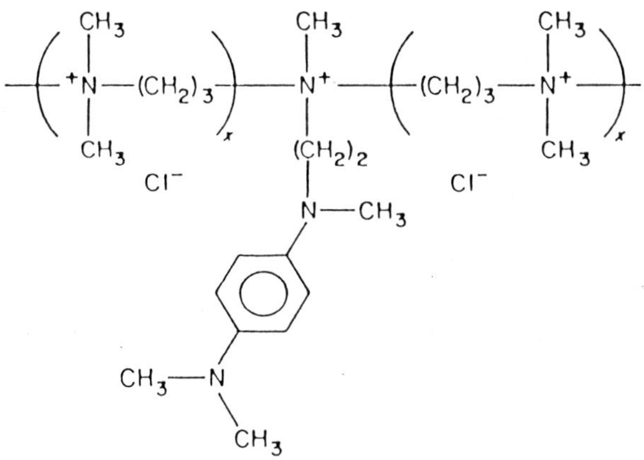

27

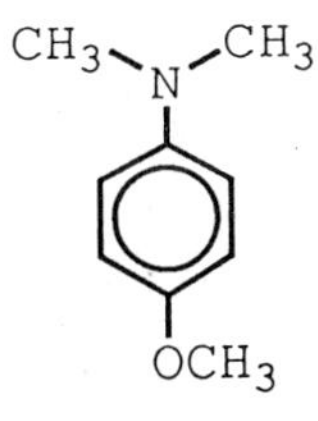

MDMA, **28**

as an electron mediator. MDMA is a neutral molecule and reacts with the photoexcited Ru(II) complex to give a reduced Ru complex and an MDMA radical cation. The MDMA radical cation oxidizes the donor moiety in **27**, thus separating the photoproducts into two different polycations.

The back ET between the reduced Ru complex in **26** and the oxidized donor moiety in **27** was strikingly slow. Figure 14 shows the decays of reduced **26** at 510 nm and oxidized **27** at 606 nm after irradiation with 450-nm light for

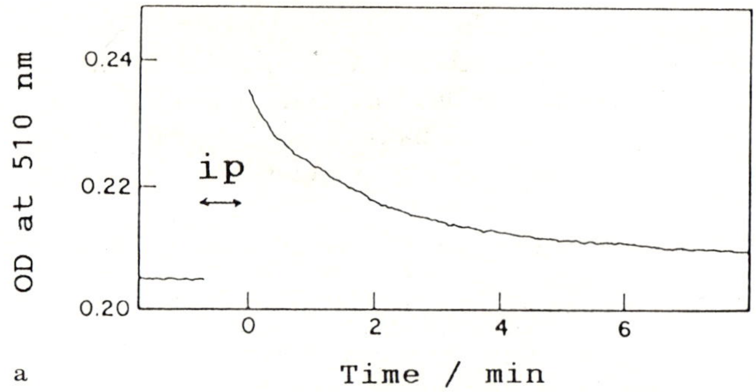

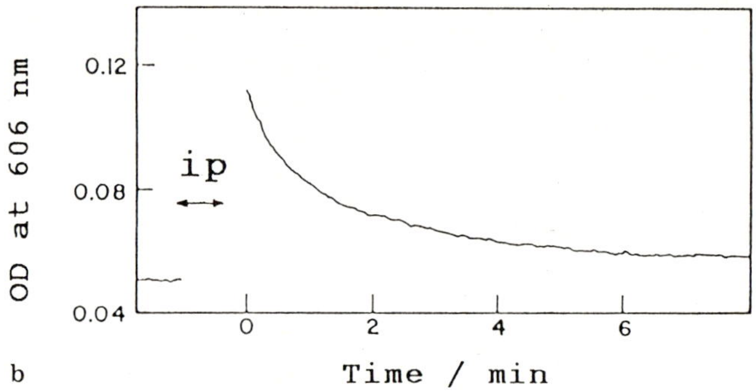

Fig. 14a, b. Decays of absorption signals of (**a**) the reduced Ru complex in **26** (510 nm) and (**b**) the oxidized donor moiety in **27** (606 nm) in the two polymer system measured after 30 s steady state irradiation at 450 nm; *ip* represents the duration of irradiation [83]

30 seconds [83]. The decays reflect second-order back ET reactions, giving $k_b = 1.5 \times 10^4\ M^{-1}\ s^{-1}$. This rate is more than five orders of magnitude slower than that for the back ET between the corresponding monomeric species.

Sassoon and Rabani [79] also prepared a two polymer system in which a chromophore was covalently bound to one polyelectrolyte and a donor or acceptor was electrostatically held by the other polyelectrolyte, and showed that its back ET underwent a similar retardation effect. They employed **26** as a photosensitizer, MV^{2+} as a mediator, and ferricyanide as an acceptor electrostatically bound to the added polycation (polybrene).

The quenching of the photoexcited Ru(II) complex in **26** by MV^{2+} occurred via a dynamic process with a rate constant of the order of $10^8\ M^{-1}\ s^{-1}$ at a low ionic strength. The back ET rate was close to the diffusion-control limit and the quantum yields of the ET fell in the range 0.05–0.11. Such relatively low net yields of charge-separated products suggest a very fast rate of the back ET occurring in the primary ion-pair state within the photochemical cage. Ferricyanide is known to be reduced by $MV^{+\cdot}$ at a virtually diffusion-controlled rate in the absence of the polyelectrolyte. Upon addition of polybrene, however, the rate was slowed by a factor of 27–40, depending on the concentration of the added polyelectrolyte. Thus, MV^{2+} receives an electron from photoexcited **26**, and $MV^{+\cdot}$ relays the electron to a ferricyanide ion electrostatically bound to polybrene. The rate constants for these two processes were of the order of $10^8\ M^{-1}\ s^{-1}$. The back ET between the photoproducts thus isolated on the different polycations was considerably slow; e.g. $k_b = 5 \times 10^7\ M^{-1}\ s^{-1}$ was estimated at a total chloride ion concentration of 0.02 M for the slowest case [79]. The idea of Rabani and Sassoon for isolating photoinduced ET products on different polyelectrolytes is very interesting and promising for improving the performance of photoinduced ET systems.

Although the electrostatic field on the polyelectrolyte surface effectively impedes back ET, it is unable to retard very fast back ET or charge recombination of the primary ion pair within the photochemical cage. The overall quantum yield of photoinduced ET is actually controlled in most cases by the charge recombination. Hence, its retardation is the key problem for attaining high quantum yields in the photoinduced ET.

6 Effects of Hydrophobic Compartmentalization of Photoinduced Electron Transfer

The microphase structure of amphiphilic polyelectrolytes in aqueous solution provides photoinduced ET with an interesting microenvironment, where a photoactive chromophore and a donor or acceptor can be held apart at different locations. Photoinduced ET in such separated donor-acceptor systems allows an efficient charge separation to be achieved.

6.1 General Considerations of Separated Donor-Acceptor Systems

Photoinduced charge separation depends, among other things, on the rates of forward and back ET, which in turn depends on such parameters as reaction exothermicity or free energy gap ($-\Delta G^0$), separation and orientation of donor (D) and acceptor (A), internal and solvent reorganization energy, and solvent polarity [84–96]. According to Marcus' classical ET theory [97–99], the ET rate increases with an increase in $-\Delta G^0$ (normal region) and reaches a maximum when $-\Delta G^0$ matches the total reorganization energy. However, it decreases at greater exothermicities owing to an increased mismatch in the Franck-Condon factors (inverted region). Thus, the ET rate shows a bell-shape dependence on free energy gap. On the basis of this theory, efficient charge separation is logically approached by choosing a combination of D and A in such a way that the back ET occurs in the inverted region with a large free energy gap for ET, while the forward ET occurs in the normal region with an appropriate driving force.

Another important factor to determine the charge separation efficiency is the distance between and the mutual orientation of the donor and the acceptor in the geminate ion-pair state. The rate of charge recombination depends on whether the geminate ion pair is "tight" or "loose" [100–102]. A "tight" ion pair, in general, has a much less chance to dissociate (charge separate) than a "loose" one. If a strong electronic interaction between D and A exists, then photoinduced ET would lead to a "tight" geminate ion-pair state with a very short life. Therefore, for efficient charge separation, D and A should be held apart to reduce their strong interaction so that the photoinduced ET can give rise to a geminate ion pair with a "loose" structure.

A number of studies have focused on D-A systems in which D and A are either embedded in a rigid matrix [103–110] or separated by a rigid spacer with covalent bonds [111–118]. Miller et al. [114, 115] gave the first experimental evidence for the bell-shape energy gap dependence in charge shift type ET reactions [114, 115]. Many studies have been reported on the photoinduced ET across the interfaces of some organized assemblies such as surfactant micelles [4] and vesicles [5], wherein some particular D and A species are expected to be separated by a phase boundary. However, owing to the dynamic nature of such interfacial systems, D and A are not always statically fixed at specific locations.

6.2 Charge Separation by Hydrophobic Compartmentalization of Chromophores

As has been described in Chapter 4, random copolymers of styrene (St) and 2-(acrylamido)-2-methylpropanesulfonic acid (AMPS) form a micelle-like microphase structure in aqueous solution [29]. The intramolecular hydrophobic aggregation of the St residues occurs when the St content in the copolymer is higher than ca. 50 mol%. When a small mole fraction of the phenanthrene (Phen) residues is covalently incorporated into such an amphiphilic polyelectrolyte, the Phen residues are hydrophobically encapsulated in the aggregate of the St residues. This kind of polymer system (poly(A/St/Phen), **29**) can be prepared by free radical terpolymerization of AMPS, St, and a small mole fraction of 9-vinylphenanthrene [119].

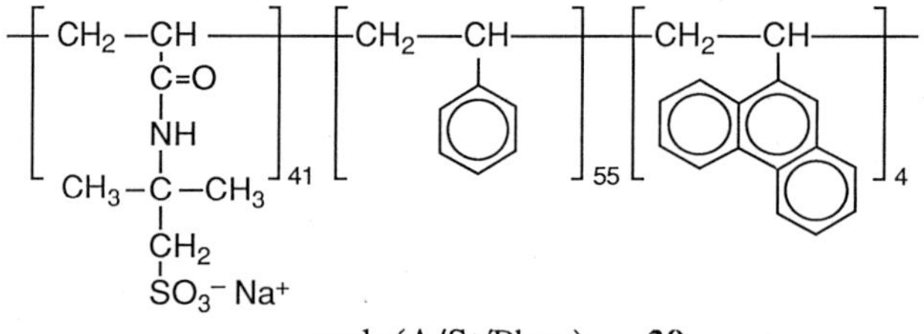

poly(A/St/Phen), **29**

Despite the fact that the Phen moieties are tightly incorporated in the "compartment" of the hydrophobic microdomain, the fluorescence from the Phen residues in poly(A/St/Phen) is very efficiently quenched by MV^{2+} in aqueous solution. The quenching efficiency is much higher than the APh-2 (**8** with x = 2)

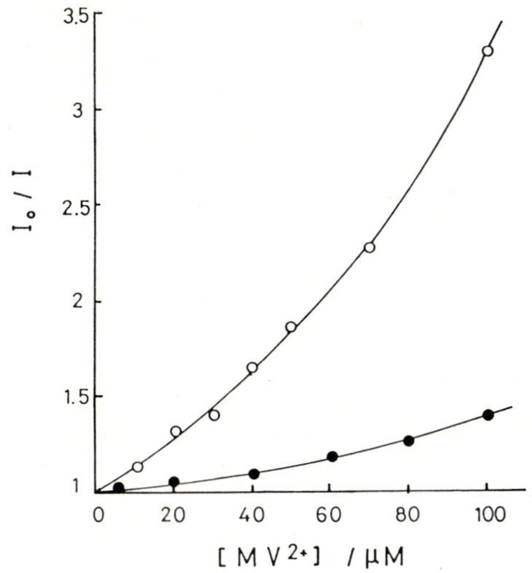

Fig. 15. Stern-Volmer plots for (○) poly(A/St/Phen) (**29**) and (●) APh-2 (**8** with x = 2) with MV^{2+} in aqueous solution; excitation wavelength, 297 nm [119]

system in which the Phen groups are not compartmentalized but exposed to the aqueous phase. The Stern-Volmer plots for the poly(A/St/Phen) and APh-2 systems are shown in Fig. 15 [119]. Being an amphiphilic dication, MV^{2+} is bound to the macromolecular environment through electrostatic and hydrophobic interactions, thus leading to a close proximity of the quencher to the Phen moieties. The steady-state quenching data do not fit to the simple Stern-Volmer kinetics, implying static quenching. In the poly(A/St/Phen) system, many MV^{2+} dications are associated with the hydrophobic aggregate of the St units by a combined electrostatic and hydrophobic interaction. The second-order equivalent rate constant of the fluorescence quenching for poly(A/St/Phen) was estimated to be $k_q = 3.8 \times 10^{11}\ M^{-1}\ s^{-1}$, which is one order of magnitude greater than the k_q value for APh-2 ($k_q = 8.9 \times 10^{10}\ M^{-1}\ s^{-1}$) [119].

As discussed in the previous chapter, the Phen residue in APh-x forms the CT complex with MV^{2+} in aqueous solution [76]. Interestingly, the CT formation is suppressed in the poly(A/St/Phen)-MV^{2+} system in spite of the Phen fluorescence being quenched by MV^{2+} very effectively. This fact indicates that it becomes very less likely for the Phen moiety to come into a face-to-face contact with MV^{2+}, while the fluorescence from the compartmentalized Phen residue can be quenched effectively via a collision-less ET to MV^{2+}.

Excitation of an aqueous solution of poly(A/St/Phen) with a 355-nm, 22-ps laser pulse in the presence of MV^{2+} generated a transient absorption band peaking at about 600 nm due to $MV^{+\cdot}$ [120]. As shown in Fig. 16, the buildup of the 600-nm band completes immediately after the pulse excitation, indicating that the photoinduced ET from the singlet-excited Phen residue ($^1Phen^*$) to MV^{2+} occurs on a time scale comparable to or shorter than the duration of the laser pulse (ca. 22 ps) [120]. Figure 16 also shows that a fast decay of the absorbance at 600 nm owing to the back ET from $MV^{+\cdot}$ to the Phen cation radical ($Phen^{+\cdot}$)

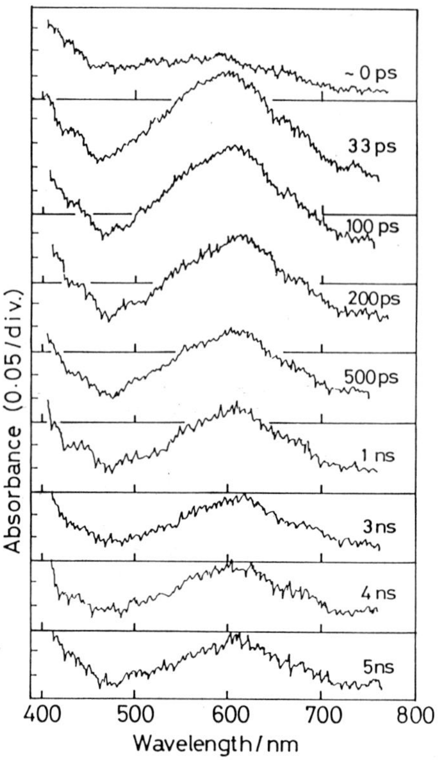

Fig. 16. Time-resolved transient absorption spectra for poly(A/St/Phen) (**29**) in aqueous solution in the presence of 5 mM MV^{2+}; [Phen] (residue) = 0.66 mM. Delay times after the laser pulse are indicated [102]

continues up to about 500 ps, after which the decay is considerably slowed. The absorbance at ca. 430 nm due to $Phen^{+\cdot}$ also decays in parallel with the 600-nm band.

For APh-2, on the other hand, the forward ET from $^1Phen^*$ to MV^{2+} was a little slower than that for the poly(A/St/Phen)-MV^{2+} system; i.e., the intensity of the $S_n \leftarrow S_1$ band for the Phen moiety at 510 nm still remained significant for 27 ps after the pulse excitation (Fig. 17) [120]. In striking contrast to the poly(A/St/Phen)-MV^{2+} system, the APh-2-MV^{2+} system showed an extremely fast decay in the transient absorbance at 600 nm over the picosecond regime and no subsequent slower decay. The transient absorbance almost completely decayed in 200 ps after the pulse.

The time profiles of the absorbance due to $MV^{+\cdot}$ at 600 nm are illustrated in Figures 18. Note that they depend on the MV^{2+} concentration. The curves for the poly(A/St/Phen)-MV^{2+} systems are biphasic and can be explained in terms of a simple mechanism illustrated in Scheme 2. Here, D‖A, A represents a compartmentalized Phen moiety (D) and MV^{2+} dications (A) bound to the hydrophobic microdomain.

The forward ET rate for the poly(A/St/Phen)-MV^{2+} system was extremely fast; i.e., the rate constant (k_{ET}) was at least of the order of 10^{11} s^{-1}. It is reasonable to consider that the fast primary electron transfer occurs from $^1D^*$ to the

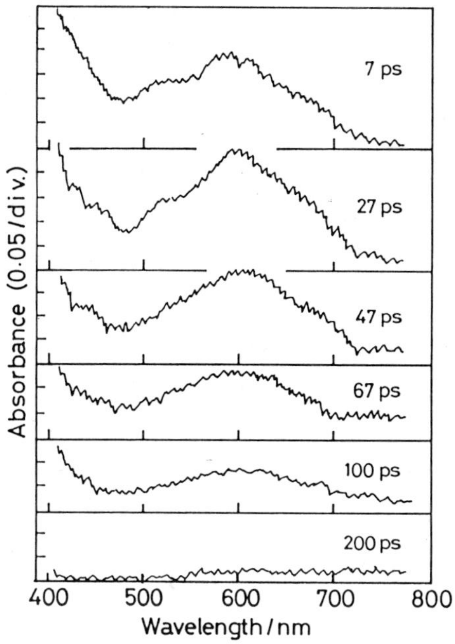

Fig. 17. Time-resolved absorption spectra for APh-2 (**8** with x = 2) in aqueous solution in the presence of 5 mM MV^{2+}; [Phen](residue) = 0.66 mM. Delay times after the laser pulse are indicated [120]

nearest-bound A (Scheme 2). The primary ion-pair state thus formed, designated $D^{+}\|A^{-}$, A in Scheme 2, may undergo a fast back ET via step (2) or a charge escape via step (3). The latter is assumed to occur via an electron exchange between $MV^{+\cdot}$ and MV^{2+} bound side-by-side to the hydrophobic microdomain as

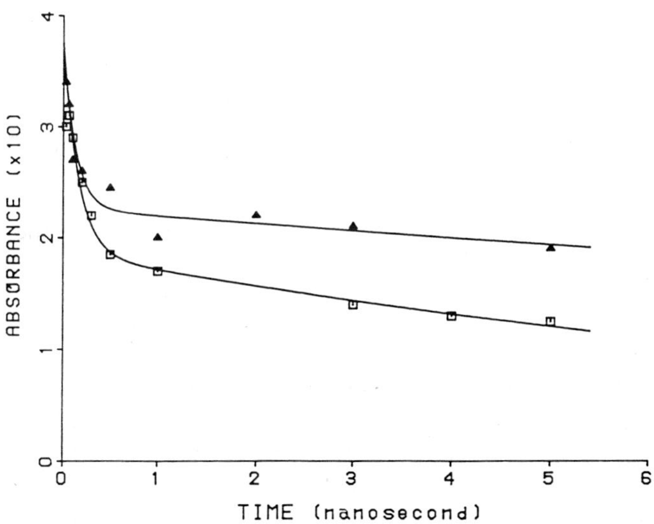

Fig. 18. Time profiles of transient absorbance at 602 nm due to $MV^{+\cdot}$ for the poly(A/St/Phen)-MV^{2+} system; [Phen](residue) = 0.66 mM; $[MV^{2+}]$ = 5 mM (□), 10 mM (△). The *solid lines* represent the best-fit curves calculated from Eq. 10 with the use of the parameters given in Table 6 [120]

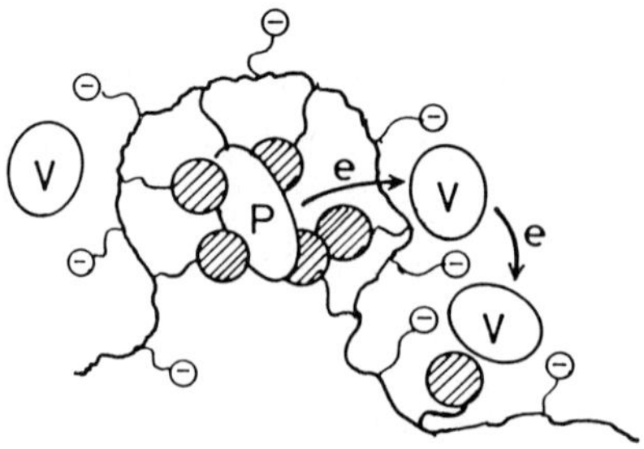

Fig. 19. Conceptual illustration of the hydrophobic compartmentalization of Phen residues (*P*) by aggregates of St residues (*shaded ovals*) and photoinduced ET followed by charge escape via an electron exchange between bound MV^{2+} (*V*) species [120]

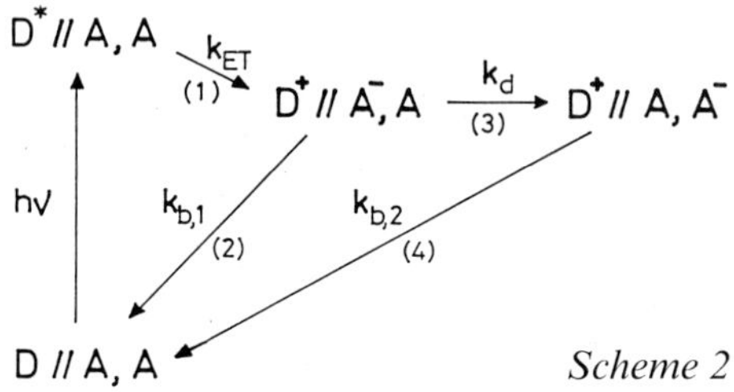

Scheme 2

conceptually illustrated in Fig. 19. Then, the back ET rate in step (4) would be considerably slowed. The transient $MV^{+\cdot}$ species observed in picosecond laser photolysis experiments are the sum of the $D^+\|A^-$, A and $D^+\|A, A^-$ species.

A mathematical treatment of the kinetic model shown in Scheme 2 gives a decay function as

$$A(t) = A(0)\,[\alpha \exp(-t/\tau_1) + (1 - \alpha)\exp(-t/\tau_2)] \tag{10}$$

with

$$\alpha = (k_{b,1} - k_{b,2})/(k_d + k_{b,1} - k_{b,2}) \tag{11}$$

$$1/\tau_1 = k_d + k_{b,1} \tag{12}$$

$$1/\tau_2 = k_{b,2} \tag{13}$$

Here, A(t) and A(0) are the 600-nm absorbances at time t and t = 0, respectively. Fit of absorbance decay data to Eq. 10 is illustrated in Fig. 18. The short-lived component in the biphasic decay is related to step (2), while the long-lived component is related to step (4). The rate constants determined by best-fitting are listed in Table 6. The charge escape yields ($\eta = k_d/(k_{b,1} + k_d)$) are also given in the table [120].

In the APh-2-MV^{2+} system, a "tight" ion pair can be formed because the motional freedom of the $Phen^{+\cdot}$ residue and a free access of $MV^{+\cdot}$ to the $Phen^{+\cdot}$ site allow the ion pair to realize an optimal distance and orientation, thus giving rise to a shorter-lived geminate ion pair. This explains why the back ET in the

Table 6. Kinetic parameters for the poly(A/St/Phen) (**29**)-MV^{2+} system [120]

$[MV^{2+}]$ mM	$k_{b,1} \times 10^{-9}$ s^{-1}	$k_d \times 10^{-7}$ s^{-1}	$k_{b,2} \times 10^{-7}$ s^{-1}	η
5	3.1 ± 1.0	3.1 ± 0.9	8.8 ± 3.2	0.50 ± 0.05
10	3.0 ± 1.1	4.7 ± 1.4	3.1 ± 2.3	0.61 ± 0.04

primary ion-pair state is very rapid in the APh-2-MV^{2+} system. The $k_{b,1}$ value for APh-2 is of the order of 10^{10} s^{-1}, being larger by one order of magnitude than that for poly(A/St/Phen). The primary ion-pair in the poly(A/St/Phen)-MV^{2+} system, on the other hand, may have a "loose" structure with a longer separation because an approach of $MV^{+\cdot}$ to $Phen^{+\cdot}$ is sterically hindered owing to the tight compartmentalization of the $Phen^{+\cdot}$ moiety. Consequently, the longer-lived ion pair can dissociate into the $MV^{+\cdot}$ and $Phen^{+\cdot}$ species with a relatively high yield.

The reaction exothermicities ($-\Delta G^0$) for forward and back ET in polar media were approximately estimated to be 1.39 and 2.18 eV, respectively [120]. Since the back ET is highly exothermic, the relatively small $k_{b,1}$ values for the compartmentalized system may be ascribed to the combined effect of the "inverted region" [97–99] and the "loose" ion-pair state.

The back ET rate for step (4) is slower than that for step (2) by two orders of magnitude. The back ET for step (4) is dependent on the MV^{2+} concentration, while that for step (2) is independent of this concentration. The k_d values are comparable to the $k_{b,1}$ values, leading to quite high charge escape yields; i.e, 50–61% depending on the MV^{2+} concentration. This is not surprising, since the rate of electron exchange between $MV^{+\cdot}$ and MV^{2+} is known to be fast and the local concentration of MV^{2+} on the surface of the microdomain is very high. Furthermore, the charge escape process (step (3)) may be assisted by the electrostatic potential field on the surface of the hydrophobic microdomain [121–123]. With increasing MV^{2+} concentration, the value of k_d increases and so does the charge escape yield (Table 6). This behavior is presumably because at a higher concentration of MV^{2+} the average distance between the bound MV^{2+} species on the domain surface is decreased. Namely, at a higher MV^{2+} concentration the $MV^{+\cdot}$ species may have more chance to give an electron to the nearest MV^{2+} species situated on the surface of the microdomain. The back ET for step (4) tends to be more retarded at a higher concentration of MV^{2+}. This tendency may contribute to further electron self-exchange steps which take place subsequent to step (3) at a higher local concentration of bound MV^{2+}. The transient absorption due to the charge-separated $MV^{+\cdot}$ species persists for hundreds of microseconds at 10 mM MV^{2+}.

The rate of ET depends on the donor-acceptor separation distance. In the poly(A/St/Phen)-MV^{2+} system, the Phen moiety is "protected" from a close contact with MV^{2+}, but the distance between the compartmentalized Phen and bound MV^{2+} species is uncertain. This means the impossibility of quantitative discussion on the ET rate in terms of the distance dependence. The spread of the distances between the donor and acceptor moieties should lead to dispersed rates of ET. Hence, the extremely large k_{ET} value observed for the compartmentalized system suggests that each Phen moiety has at least one MV^{2+} so close to its site that there occurs a fast forward ET.

Delaire et al. [124] have reported that laser photolysis of an acidic solution (pH 2.8) containing PMAvDPA (**23**) and MV^{2+} allows the formation of surprisingly long-lived $MV^{+\cdot}$ and DPA cation radicals with a very high charge escape quantum yield. The content of the DPA chromophores in PMAvDPA is as low as less than 1/1000 in the molar ratio DPA/MAA. Figure 20 shows a decay profile of the transient absorption due to $MV^{+\cdot}$ monitored at 610 nm [124]. The absorption persists for several milliseconds. As depicted in Fig. 20, the decay obeys second-order kinetics, which yields $k_b = 3.5 \times 10^8\ M^{-1}\ s^{-1}$. From the initial optical density measured at 610 nm, the quantum yield for charge escape was estimated to be 0.72 at 0.2 M MV^{2+}.

It has been shown in Chapter 5, the fluorescence quenching of the DPA moiety by MV^{2+} is very efficient in an alkaline solution [60]. On the other hand, Delaire et al. [124] showed that the quenching in an acidic solution (pH 1.5–3.0) was less effective ($k_q = 2.5 \times 10^9\ M^{-1}\ s^{-1}$); i.e., it was slower than the diffusion-controlled limit. They interpreted this finding as due to the reduced accessibility of the quencher to the DPA group located in the hydrophobic domain of protonated PMA at acidic pH. An important observation is that, in a basic medium, laser excitation of the PMAvDPA-MV^{2+} system yielded no transient absorption. This implies that a rapid back ET occurs after very efficient fluorescence quenching.

More recently, Webber et al. [125, 126] have extensively studied photoinduced ET for various hydrophobic chromophores covalently bound to PMA by using

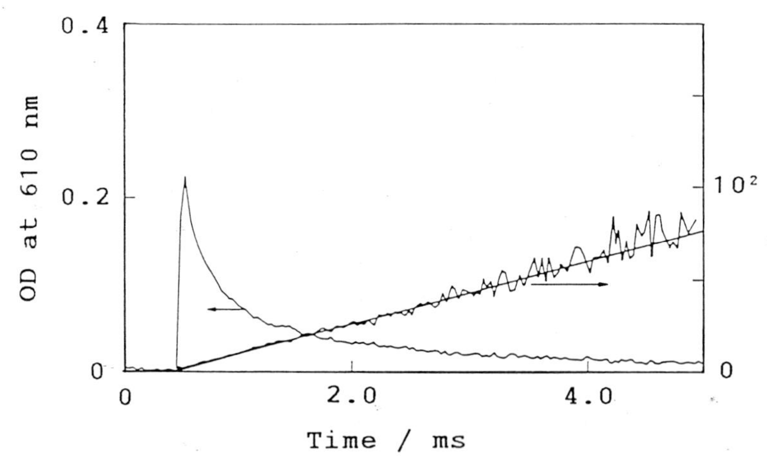

Fig. 20. Transient decay of absorption at 610 nm due to $MV^{+\cdot}$ after laser pulse excitation of an acidic aqueous solution (pH = 2.8) of PMAvPy (**23**) in the presence of 0.2 M MV^{2+} [124]

MV^{2+} or SPV as an acceptor [125, 126]. They have concluded that the "steric protection" of chromophores from the quencher crucially affects the efficiency of the initial charge separation. This conclusion is essentially the same as that reached by Morishima et al. [119, 120].

The effect of the "steric protection" on the initial charge separation varies with the kind of chromophore. For the pyrene moiety in PMAvPy (**20**), no charge separation was observed [77], although there was evidence for some degree of steric protection by the compact PMA coil at acidic pH, in striking contrast to the DPA system. It should be noted that PMAvPy shows much stronger static quenching than does PMAvDPA even at low pH, which implies that the primary ion-pair state in the pyrene system has a "tight" structure that allows rapid charge recombination to occur. Webber et al. [77] have speculated that the steric protection of the Py moiety by the protonated PMA coil is less complete than that of the DPA moiety and that the phenyl groups in the 9 and 10 positions of the DPA group may hinder the approach of the quencher to the anthryl ring.

The experimental results on poly(methacrylic acid) containing a small mole fraction of either 3-vinylperylene (PMAvPER, (**30**)) or *N*-[12-(4-aminonaphthalimide)]-2-methylacrylamide (PMAANI, (**31**)) show charge separation which is efficient for PMAvPER but not much for PMAANI. The quantum yields of charge separation for various chromophores covalently bound to PMA at pH 2.8 are summarized in Table 7.

From the relation between the quenching efficiency and the charge separation efficiency Webber et al. [77] and Morishima et al. [76, 119, 120] have reached the same conclusion that efficient fluorescence quenching static in nature does not lead to efficient charge separation. This conclusion seems to apply generally to

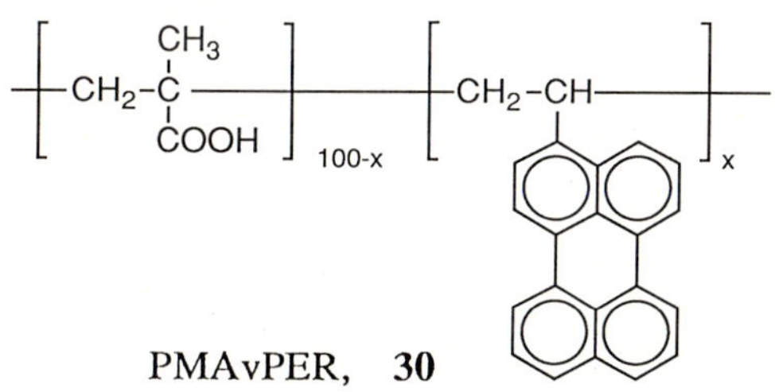

PMAvPER, **30**

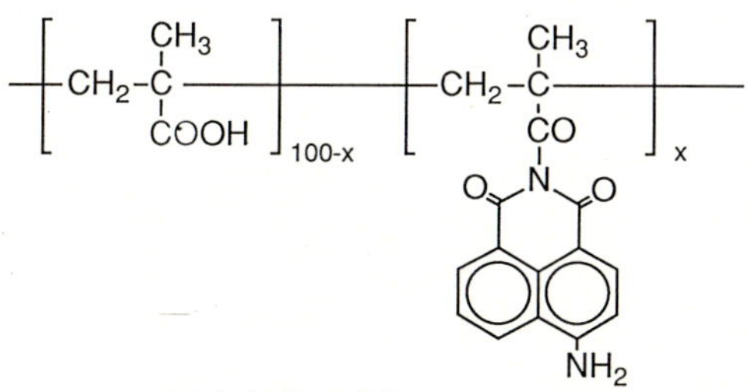

PMAANI, **31**

Table 7. Quantum yields of charge separation, φ_{cs}, for poly(methacrylic acid)-bound chromophores at pH 2.8 [77]

Polymer	$[MV^{2+}]$/mM	φ_{cs}
PMAvDPA (**23**)	50	0.73
	100	0.75
	200	0.72
PMAvPER (**30**)	10	0.35
	38.7	0.36
	75	0.36
PMAANI (**31**)	10	0.50
	38.7	0.19
	75	0.11
PMAvPy (**20**)	no charge separation	

photoinduced ET systems in which acceptors (or donors) can freely come close to photoexcited chromophores.

The amphiphilic polyelectrolyte systems discussed in this chapter provide an unusual microenvironment in which close approach of the acceptor to the photoexcited chromophore site is sterically hindered so that efficient charge separation is made possible. Furthermore, the compartmentalization of the chromophore reported by Morishima et al. [119, 120] is unique in that, differing from the general conclusion made above, both efficient "static" quenching and charge separation occur. The key idea for the design of photoinduced ET systems with efficient quenching and charge separation is how to keep D and A apart at an optimal distance to prevent the ground-state CT complexation from occurring, while a rapid ET occurs to generate a "loose" ion pair.

7 Concluding Remarks

Although the electrostatic potential on the surface of the polyelectrolyte effectively prevents the diffusional back electron transfer, it is unable to retard the very fast charge recombination of a geminate ion pair formed in the primary process within the photochemical cage. Compartmentalization of a photoactive chromophore in the microphase structure of the amphiphilic polyelectrolyte provides a separated donor-acceptor system, in which the charge recombination is effectively suppressed. Thus, with a compartmentalized system, it is possible to achieve efficient charge separation.

Besides the photoinduced electron transfer reaction discussed in the present review, the compartmentalized chromophore system may provide many other interesting photo-systems worthy of studying in their own interest. For example, energy migration and transfer within the compartment may be the keys to design an interesting photon harvesting system [11], which is often referred to as the "antenna" system [127]. If a hydrophobic chromophore with a lower photoexcitation energy is compartmentalized by an aggregate of different chromophores with a higher energy, the excitation energy, which migrates through the chromophore aggregate [30], can be trapped by the compartmentalized chromophore. Thus, the photon energy can be collected on the specific chromophore. It is worth noting here that, very recently, Nowakowska et al. [128] have shown a novel antenna polyelectrolyte which behaves as an efficient photocatalyst and so is referred to as a "photozyme".

The excited singlet and triplet states of a compartmentalized chromophore are markedly longer lived than those of a free chromophore, owing to the isolation as well as tight fixation effects. It appears that the isolated chromophore system fulfills some critical requirements for a photochemical hole burning (PHB) system. Therefore, applying compartmentalized systems to the PHB system may be an interesting challenge.

Since the compartmentalization occurs as a result of microphase separation of an amphiphilic polyelectrolyte in aqueous solution, an aqueous system is the only possible object of study. This limitation is a disadvantage from a practical point of view. Our recent studies, however, have shown that this disadvantage can be overcome with a molecular composite of an amphiphilic polyelectrolyte with a surfactant molecule [129]. This composite was dissolvable in organic solvents and dopable in polymer film, and the microphase structure was found to remain unchaged in the composite. This finding is important, because it has made it possible to extend the study on photo-systems involving the chromophore compartmentalization to organic solutions and polymer solid systems.

Acknowledgement: The author expresses his appreciation to Professor Hiroshi Fujita for giving him the opportunity of writing this article and for editing the manuscript. He is also grateful to Professor Mikiharu Kamachi for his valuable discussion and continuing encouragement.

8 References

1. Bolton JR (1977) Solar power and fuel. Academic, New York
2. Hautala RR, King RB, Kutal C (1979) Solar energy, chemical conversion and storage. The Human Press, Clinto NJ
3. Barber J (1979) Photosynthesis in relation to model systems. Elsevier, New York
4. Thomas JK (1984) The chemistry of excitation at interfaces. ACS Monograph Series, No 181, American Chemical Society, Washington, D.C.
5. Fendler J (1983) Membrane mimetic chemistry. Academic press, New York
6. Wilner J, Ford WE, Otvos JW, Calvin M (1979) Nature (London) 280: 823
7. Rodgers MAJ, Becker JC (1980) J Phys Chem 84: 2762
8. Jones CA, Weaner LE, Mackay RA (1980) J Phys Chem 84: 1495
9. Kalyanasundaram K (1987) Photochemistry in microheterogeneous systems. Academic, New York
10. Rabani J (1988) In: Fox MA, Chanon M (eds) Photoinduced electron transfer. Elsevier, Amsterdam
11. Morishima Y (1990) Prog Polym Sci 15: 949
12. Morawetz H (1970) Acc Chem Res 3: 354
13. Okubo T, Ise N (1972) Proc R Soc London Ser A 327: 413
14. Okuba T, Ise N (1973) Bull Chem Soc Jpn 46: 2493

15. Mita K, Kunugi S, Okubo T, Ise N (1975) J Chem Soc Faraday Trans 1 71: 936
16. Okubo T, Ishiwatari T, Mita K, Ise N (1975) J Phys Chem 79: 2108
17. Ise N, Okubo T (1978) Macromolecules 11: 439
18. Meisel D, Matheson MS (1977) J Am Chem Soc 99: 6577
19. Meisel D, Rabani J, Meyerstein D, Matheson MS (1978) J Phys Chem 82: 985
20. Jonah CD, Matheson MS, Meisel D (1979) J Phys Chem 83: 257
21. Meyerstein D, Rabani J, Matheson MS, Meisel D (1978) J Phys Chem 82: 1879
22. Sassoon RE, Rabani J (1980) J Phys Chem 84: 1319
23. Ladenheim H, Morawetz H (1959) J Am Chem Soc 81: 20
24. Lovrien R, Waddington JC (1964) J Am Chem Soc 86: 2315
25. Strauss UP, Jackson EG (1951) J Polym Sci 6: 649
26. Strauss UP, Gershfeld NL, Crook EH (1956) J Phys Chem 60: 577
27. Dubin P, Strauss UP (1967) J Phys Chem 71: 2757
28. Dubin P, Strauss UP (1970) J Phys Chem 74: 2842
29. Morishima Y, Itoh Y, Nozakura S (1981) Makromol Chem 182: 3135
30. Morishima Y, Kobayashi T, Nozakura S (1985) J Phys Chem 89: 4081
31. Morishima Y, Kobayashi T, Nozakura S (1989) Polym J 21: 267.
32. Morawetz H (1966) Macromolecules in solution. Interscience, New York, p 348
33. Fuoss RM, Katchalsky A, Lifson S (1951) Proc Natl Acad Sci USA 37: 579
34. Alexandrowicz Z, Katchalsky A (1963) J Polym Sci Polym Phys Ed 1: 3231
35. MacGillivray AD, Winklemann Jr JJ (1966) J Chem Phys 45: 2184
36. Dolar D, Peterlin A (1969) J Chem Phys 50: 3011
37. Stigter D (1975) J Colloid Interf Sci 53: 296
38. Lebret M, Zimm BH (1984) Biopolymers 23: 287
39. Gueron M, Weisbuch G (1980) Biopolymers 19: 353
40. Drummond CJ, Grieser F, Healy TW (1988) J Phys Chem 92: 2604
41. Steiner U, Abdel-Kader MH, Fischer P, Kramer HEA (1978) J Am Chem Soc 100: 3190
42. Morishima Y, Higuchi Y, Kamachi M (1991) J Polym Sci Polym Chem Ed 29: 677
43. Manning GS (1965) J Chem Phys 43: 4250
44. Manning GS (1979) Acct Chem Res 12: 443
45. Manning GS (1974) In: Selegny E (ed) Polyelectrolytes. Reidel Dordrecht, The Netherlands, p 9
46. Hartly GS, Roe JW (1940) Trans Faraday Soc 36: 101
47. Mukerjee P, Banerjeem K (1964) J Phys Chem 68: 3567
48. Fernàndez MS, Fromherz P (1977) J Phys Chem 81: 1755
49. Morishima Y, Kobayashi T, Nozakura S (1988) Macromolecules 21: 101
50. Itoh Y, Morishima Y, Nozakura S (1982) J Polym Sci Polym Chem Ed 20: 467
51. Morishima Y, Itoh Y, Hashimoto T, Nozakura S (1982) J Polym Sci Polym Chem Ed 20: 2007
52. Morishima Y, Tanaka T, Itoh Y, Nozakura S (1982) Polym J 14: 861
53. Morishima Y, Itoh Y, Nozakura S (1982) Chem Phys Lett 90: 258
54. Morishima Y, Nozakura S (1986) J Polym Sci Polym Symp 74: 1
55. Morishima Y, Kobayashi T, Nozakura S, Webber SE (1987) Macromolecules 20: 807
56. Morishima Y, Lim HS, Nozakura S, Sturtevant JL (1989) Macromolecules 22: 1148
57. Guillet JE, Takahashi Y, McIntosh AR, Bolton JR (1985) Macromolecules 18: 1788
58. Guillet JE, Rendall WA (1986) Macromolecules 19: 224
59. Guillet JE, Wang J, Lu L (1986) Macromolecules 19: 2793
60. Delaire JA, Rodgers MAJ, Webber SE (1984) J Phys Chem 88: 6219
61. Bai F, Chang C-H, Webber SE (1986) Macromolecules 19: 588
62. Turro NJ, Arora KS (1986) Polymer 27: 783
63. Arora KS, Hwang K-C, Turro NJ (1986) Macromolecules 19: 2806
64. Chu C-Y, Thomas JK (1984) Macromolecules 17: 2142
65. Anson FC, Saveant J-M, Shigehara K (1983) J Am Chem Soc 105: 1096
66. Anson FC, Ohsaka T, Saveant J-M (1983) J Phys Chem 87: 640
67. Montgemery DD, Anson FC (1985) J Am Chem Soc 107: 3431
68. Sbiti N, Tondre C (1984) Macromolecules 17: 369

69. Jager J, Engberts JBFN (1985) J Org Chem 50: 1474
70. Nemethy G, Scheraga HA (1962) J Chem Phys 36: 3382
71. Nemethy G, Scheraga HA (1962) J Phys Chem 66: 1773
72. Jencks WP (1969) Catalysts in chemistry and enzymology. McGraw-Hill, New York, p 393
73. Itoh Y, Morishima Y, Nozakura S (1983) J Polym Sci Polym Lett Ed 21: 167
74. Itoh Y, Morishima Y, Nozakura S (1984) Photochem Photobiol 39: 451
75. Itoh Y, Morishima Y, Nozakura S (1984) Photochem Photobiol 39: 603
76. Morishima Y, Itoh Y, Nozakurza S, Ohno T, Kato S (1984) Macromolecules 17: 2264
77. Stramel RD, Ngyen C, Webber SE, Rodgers MAJ (1988) J Phys Chem 92: 2934
78. Webber SE (1986) Macromolecules 19: 1658
79. Sassoon RE, Rabani J (1985) J Phys Chem 89: 5500
80. Kaneko M, Yamada A, Tsuchida E, Kurimura Y (1984) J Phys Chem 88: 1061
81. Kaneko M, Nakamura H (1987) Macromolecules 20: 2265
82. Ennis PM, Kelly JM, O'Connell CM (1986) J Chem Soc Dalton Trans 2485
83. Sassoon RE (1985) J Am Chem Soc 107: 6133
84. Strauch S, McLendon G, McGuire M, Guarr T (1983) J Phys Chem 87: 3579
85. Joran AD, Leland BA, Geller GG, Hopfield JJ, Dervan PB (1984) J Am Chem Soc 106: 6090
86. Leland BA, Joran AD, Felker PM, Hopfield JJ, Zewail AH, Dervan PB (1985) J Phys Chem 89: 5571
87. Joran AD, Leland BA, Felker PM, Zewail AH, Hopfield JJ, Dervan PB (1987) Nature 327: 508
88. Miller JR, Beitz JV, Huddleston RK (1984) J Am Chem Soc 106: 5057
89. McLendon G, Miller JR (1985) J Am Chem Soc 107: 7811
90. Gould IR, Ege D, Mattes SL, Farid S (1987) J Am Chem Soc 109: 3794
91. Hwang J-K, Warshel A (1987) J Am Chem Soc 109: 715
92. Isied SS, Vassilian A, Wishart JF, Creutz C, Schwarz HA, Sutin N (1988) J Am Chem Soc 110: 635
93. Kakitani T, Mataga N (1985) Chem Phys p 381
94. Kakitani T, Mataga N (1985) J Phys Chem 89: 4752
95. Kakitani T, Mataga N (1986) J Phys Chem 90: 993
96. Yoshimori A, Kakitani T, Enomoto Y, Mataga N (1989) J Phys Chem 93: 8316
97. Marcus RA (1960) Discuss Faraday Soc 29: 21
98. Marcus RA (1985) J Chem Phys 43: 2654
99. Marcus RA, Siders P (1982) J Phys Chem 86: 622
100. Mataga N, Kanda Y, Okada T (1986) J Phys Chem 90. 3880
101. Mataga N, Asahi T, Kanda Y, Okada T, Kakitani T (1988) Chem Phys 127: 249
102. Asahi T, Mataga N (1989) J Phys Chem 93: 6575
103. Miller JR (1975) Science (Washington, D.C.) 189: 221
104. Beitz JV, Miller JR (1979) J Phys Chem 71: 4579
105. Miller JR, Beitz JV (1981) J Phys Chem 74: 6746
106. Miller JR, Peaples JA, Schmitt MJ, Closs GL (1982) J Am Chem Soc 104: 6488
107. Strauch S, McLendon G, McGuire M, Guarr T (1983) J Phys Chem 87: 3579
108. Guarr T, McGuire M, Strauch S, McLendon G (1983) J Am Chem Soc 105: 616
109. Milosavljevic BH, Thomas JK (1985) Chem Phys Lett 114: 133
110. Milosavljevic BH, Thomas JK (1986) J Am Chem Soc 108: 2513
111. Stein CA, Lewis NA, Seitz G (1982) J Am Chem Soc 104: 2596
112. Pasman P, Koper NW, Verhoeven JW (1982) Recl Trav Chim Pays-Bas 101: 363
113. Mes GF, Van Ramesdonk HJ, Verhoeven JW (1984) J Am Chem Soc 106: 1335
114. Calcaterra LT, Closs GL, Miller JR (1983) J Am Chem Soc 105: 670
115. Miller JR, Calcaterra LT, Closs GL (1984) J Am Chem Soc 106: 3047
116. Wasielewski MR, Niemczyk MP (1984) J Am Chem Soc 106: 5043
117. Wasielewski MR, Niemczyk MP, Svec WA, Pewitt EB (1985) J Am Chem Soc 107: 1080
118. Beratan DN (1986) J Am Chem Soc 108: 4321
119. Morishima Y, Kobayashi T, Furui T, Nozakura S (1987) Macromolecules 20: 1707

120. Morishima Y, Furui T, Nozakura S, Okada T, Mataga N (1989) J Phys Chem 93: 1643
121. Takuma K, Sakamoto T, Nagamura T, Matsuo T (1981) J Phys Chem 85: 619
122. Matsuo T, Sakamoto T, Takuma K, Sakura K, Ohsaka T (1981) J Phys Chem 85: 1277
123. Matsuo T (1982) Pure Appl Chem 54: 1693
124. Delaire JA, Rodgers MAJ, Webber SE (1986) Eur Polym J 22: 189
125. Delaire JA, Sanquer-Barrie M, Webber SE (1988) J Phys Chem 92: 1252
126. Stramel RD, Webber SE, Rodgers MAJ (1988) J Phys Chem 92; 6625
127. Guillet J (1985) In: Polymer photophysics and photochemistry. Cambridge University Press, London, p 241
128. Nowakowska M, Sustar E, Guillet JE (1991) J Am Chem Soc 113: 253 and references therein
129. Morishima Y, Seki M, Tominaga Y, Kamachi M (to be published) Macromolecules

Editor: H. Fujita
Received March 21, 1991

Chemistry and Physics of "Agricultural" Hydrogels

K. S. Kazanskii and S. A. Dubrovskii
Institute of Chemical Physics, Academy of Sciences of the USSR, Moscow V-334, USSR

Superabsorbent polymer hydrogels can swell to absorb huge volumes of water or aqueous solutions. This property has led to many practical applications of these new materials, in particular, in agriculture for improving water retention of soils and the water supply of plants. This article reviews methods of superabsorbent gel synthesis, measurements and treatment of their properties, as well as their effects in soil and on plant growth. The thermodynamic approach used to describe the swelling behavior of polymer networks proves to be quite helpful in modelling the hydrogel efficiency as a water-absorbing additive.

List of Abbreviations

AAc	Acrylic acid
AAm	Acrylamide
AN	Acrylonitrile
MBAA	*N*, *N'*-Methylene-bis-acrylamide
PAAc	Poly(acrylic acid)
PAAm	Polyacrylamide
PAN	Polyacrylonitrile
PEG	Poly(ethylene glycol)
PEO	Poly(ethylene oxide)
PVA	Poly(vinyl alcohol)
SAH	Superabsorbent hydrogel

1 Introduction

Hydrogels are three-dimensionally crosslinked hydrophilic polymers capable of swelling in water and retaining possibly huge volumes of water in the swollen state. These peculiar materials are of great interest due to their promising applications in medicine, in solving some ecological and biological problems, as well as in modern technologies.

The first generation of hydrogels with swelling up to 40–50% appeared in the late 1950s and was mainly based on hydroxyalkyl methacrylate and related monomers. These hydrogels were used in developing contact lenses which marked a revolution in ophthalmology [1]. The need to improve these lenses, and the emerging number of other medical problems [2, 3] resulted in the development of a second generation of hydrogels with swelling degrees of 70–80%, which widened the scope of their application.

From a theoretical viewpoint [4], the equilibrium swelling of polymer networks is attained with the balance of osmotic forces determined by the affinity of macromolecules to the solvent and their elasticity, which is limited by the presence of crosslinks. Evaluations have shown that hydrogels with much higher swelling degrees are obtainable: an isolated network chain in the form of a Gaussian coil with chain length x_c varying from 10^2 to 10^4 will have a swelling degree in a θ-solvent equal to $w \sim x_c^{1/2}$, i.e, on the order of 10–100 $g \cdot g^{-1}$. The same network in a good solvent ($w \sim x_c^{4/5}$) can swell as much as 40–1600 times. An additional factor contributing to swelling is the charged state of the chain which is obtainted by incorporating ionogenic groups of various kinds. The molecular design of highly swellable hydrogels is based on the following three parameters: affinity to the solvent, crosslinking density and ionization degree. It should be noted that it was in hydrogels where swelling as an exclusive property of polymers was practically applied for the first time, while earlier it was mainly used in network characterization.

An intensive development of hydrogels with ultra-high absorbing power began in the late 1970s to early 1980s thanks to their expanding application. Some of these uses are various sanitary tools with high absorption of physiological fluids, materials for treating burns and open wounds, nutrient media for microorganisms, moisture absorbing and anti-condensing coatings, materials for liquid waste treatment, etc. [5]. Also of importance is the possibility to use hydrogels in passive solar energy systems (the so-called "solar ponds") [6].

However, the greatest attention was given to the application of superabsorbent polymer hydrogels (SAH) in agriculture, soil improvement, and plant growth. SAH particles distributed in soil are capable of absorbing water, as a result this should efficiently improve the water-holding capacity of the soil and promote optimal plant growth. Numerous investigations confirm the soundness of this idea.

The very first studies with radiation crosslinked polyethylene oxide (PEO) have shown that SAH is able to substantially reduce the sensitivity of plants to water shortage [7], to promote their growth, particularly, under conditions of water deficiency [8], to improve seedling survival and the final crop [9]. These results stimulated a more detailed analysis of the effects of SAH in the water balance of soils and in plant growth. For example, it was discovered that an increase in water retention in sandy soils by hydrogels is due to an apparent shift of the size distribution of soil pores towards water-retaining capillaries, on the one hand, and to a decrease in physical evaporation of water, on the other [10, 11]. These two factors are quite sufficient to considerably increase the degree of water utilization by plants and, as a result, to promote more frequent and faster germination, to improve biological activity, efficiency of mineral nutrition, and finally, to increase the biomass production [12, 13].

Besides a direct addition to growth media, there are other ways of utilizing SAH: for example, coating of seeds and bare roots, drilling of germinated seeds in a swollen gel ("fluid drilling" [14, 15]), plant nutrient media for hydroponics, etc.

Therefore, polymers can now be applied to problems such as soil erosion, the consequences of droughts, food supply, and life support in arid zones.

The development of water-swellable polymers depends on aspects of their synthesis, properties evaluation, optimization and correlation of these properties with synthesis conditions. Obviously, studying the behavior of SAH in contact with liquid and solid phases of the soil as well as with plants requires developing physical models and algorithms suitable for the prediction of SAH efficiency.

This set of problems is dealt with in the present review which seems to be the first attempt to analyze this new field in the chemistry and physics of polymers.

2 Synthesis of Superabsorbent Hydrogels

Crosslinked polymers with high equilibrium swelling in water or aqueous solutions can be based exclusively on macromolecules with high hydrophilicity and flexibility, often in combination with the polyelectrolyte nature of chains. The method for SAH synthesis must provide low crosslink density of the resultant polymer network. The starting materials are either monomers (polymerization, copolymerization) or oligomers (polycondensation, polyaddition), polymers (crosslinking, modification), as well as various combinations of these materials (grafting, co-crosslinking).

The most promising trends in creating effective "agricultural" SAH are associated with polyacrylamide (PAAm) hydrogels containing a certain amount of ionogenic groups most often in the form of acrylic acid units (AAc) as well as with networks based on poly(acrylic acid) (PAAc), poly(methacrylic acid), or their alkali metal salts. Among other polymers, some polyacrylates are of interest, e.g., with amino groups, polyamines, PEO, and poly(vinyl alcohol) (PVA) gels. Natural polymers like cellulose, starch, lignin, dextran, alginates, etc. do not form hydrogels with very high water-absorbent capacity and require modification or combination with hydrophilic synthetic polymers.

2.1 Three-Dimensional Polymerization

Three-dimensional polymerization appears to be the most universal and widely used technique for obtaining SAH. The network formation proceeds either with participation of a multifunctional branching agent or due to side reactions, e.g., chain transfer to polymer. The first procedure is preferable since it facilitates the regulation of the crosslink density. Polymerization can be initiated using initiators or induced by physical factors.

2.1.1 Polymerization of Acrylic Monomers

Monomers of the acrylic series exhibit high reactivity in free-radical polymerization, many of them being capable of forming sufficiently hydrophilic polymers. For this reason various acrylates form the principal monomeric basis of hydrogels.

The acrylamide (AAm) – *N,N′*-methylene-bis-acrylamide (MBAA) system described in a great number of papers is most suitable for SAH synthesis. For example, this system is widely used in obtaining hydrogels for electrophoresis [16]. Persulfate initiators in combination with activating additives provide a high rate of initiation at 20–50 °C, and usually the formation of macroscopic gel is achieved within tens of minutes, though the system is kept for some more time to complete the monomer conversion [17]. The simplicity of preparation and the possibility of changing the network density by varying the MBAA fraction are rather attractive. Therefore, hydrogels of this particular type are most widely used as models in physical research (see Sect. 3).

The data on the composition and swelling of PAAm hydrogels are given in Table 1. It can be seen that swelling of nonionic hydrogels lies in the range of 20–250 ml · g^{-1} growing with a decrease of the MBAA fraction and the total concentration of the monomers. Ionogenic groups which abruptly increase the swelling degree are introduced into the network either by direct copolymerization of AAm with AAc and its salts or by partial hydrolysis in the alkali medium [19,

Table 1. Swelling of polyacrylamide hydrogels obtained by free-radical crosslinking polymerization

Conditions of synthesis			Swelling[2],	Refs.
Conc. of AAm, wt%	Fraction of MBAA, wt%	Content of AAc units[1], mol%	ml · g^{-1}	
6.7	0.66	–	76[a]	[18]
6.7	2.6	–	32[a]	[18]
6.7	14.0	–	22[a]	[18]
5.0	2.66	–	35[a]	[19]
5.0	2.66	2.4[a]	276[a]	[19]
3.0	1.33	–	111[a]	[20]
15.0	1.33	–	12,3[a]	[20]
15.0	1.25	40[b]	100[a]	[21]
15.0	2.5	40[b]	80[a]	[21]
10.0	0.01	5.0[a]	10000[a]; 230[b]	[22]
10.0	0.02	5.0[a]	6700[a]; 160[b]	[22]
10.0	0.05	5.0[a]	3300[a]; 57[b]	[22]

[1] Method of introducing: [a] copolymerization; [b] hydrolysis.
[2] [a] in water; [b] in 0.15 N aqueous NaCl

21, 22]. Data on similar linear copolymers show [23] that in the first case the structure is predominantly sequential, while upon hydrolysis it is more random. The swelling of hydrogels obtained by these two techniques is also different [19, 21]. However, the experimental data available are insufficient to make unambiguous conclusions on the effect of the bound charges distribution on swelling. Nevertheless, changes in the ionization degree of the network fragments make it possible to vary the swelling within wide limits, which gives a powerful tool to the chemists.

The crosslink density, another very important structural parameter of the network, is varied by changing the fraction of the crosslinking agent in the monomer mixture. Analysis of this possibility in a wider range of AAm and MBAA concentrations [18, 22, 24] points to a significant difference between the ideal and real network densities, which may be due to its topological imperfection and other circumstances. Thus, for example, the value of M_c calculated from the mixture composition and calculated from the measured elastic modulus are close to each other only in concentrated hydrogels with very low MBAA fraction (Fig. 1). It should be mentioned that this situation is generally characteristic of free-radical polymerization in many similar systems.

Serious deviations of the polymer network structure from the ideal one can have several causes. One of them is the crosslinking agent involvement in intramolecular cycle formation. The contribution of this reaction grows with the system dilution as well as when the crosslinker units in the chain are close one to the other, i.e. its fraction in the copolymer increases. All this is in good agreement with the observed trend.

The second reason is of kinetic nature and due to a substantially lower reactivity of the pendant acrylic groups. Some experimental facts may be considered a direct evidence of this possibility. Thus, the absorption in the region of 280 nm in the UV spectrum points to a growth of the residual concentration of acrylamide groups with an increasing fraction of the crosslinker [25]. The same groups can

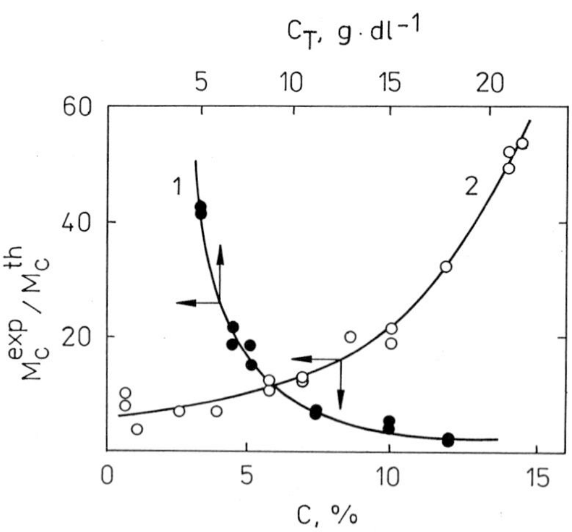

Fig. 1. Comparison of experimental and theoretical values of M_c at free-radical copolymerization of AAm with MBAA as a crosslinking agent; C_T – total concentration of monomers, C – that of MBAA; C = 10 wt% (*1*), C_T = 6.7 g · dl^{-1} (*2*). From Baselga et al. [18]

be responsible for the absorption band at $1290 \, cm^{-1}$ in the Raman spectrum of hydrogels [26].

Finally, one more reason consists in the evolution of a distinct heterogeneous structure at the formation of hydrogels in the region of high MBAA concentrations. It is known [27–31] that homogeneous hydrogels are formed only when the MBAA content is not higher than 2–4%. Structural segregation established through direct [27–29] or indirect [24, 30, 31] observations is usually described in terms of the so-called "two-phase" model in which strongly crosslinked nuclei are surrounded by much looser shells.

A recent series of papers [18, 24, 32–34] substantially clears up the three-dimensional polymerization mechanism in the AAm-MBAA system. Direct observation of the various types of acrylamide group consumption using NMR technique, analysis of conversion at the gel-point, and correlation of the elastic modulus with swelling indicate a considerable deviation of the system from the ideal model and a low efficiency of MBAA as a crosslinker. Most of these experimental data, however, refer to the range of heterogeneous hydrogels where swelling is not more than $80 \, ml \cdot ml^{-1}$ [24].

Summing up the above, let us note that compositions for obtaining SAH in three-dimensional polymerization of AAm should be chosen from the range of extremely low concentrations of the crosslinking agent. As an evidence, we have synthesized [22] a series of weakly crosslinked hydrogels (their characteristics are also given in Table 1); the network density of these SAH were found to be close to the theoretical one. Further optimization is possible when taking into account all the requirements to swelling, elastic modulus, and the ionic group content.

Hydrogels based on PAAm with a preset heterogeneous structure are obtained under the conditions of cryostructurization of the reaction system, i.e. in frozen solutions [35, 36]. At temperatures below 0 °C the reaction proceeds only in intercrystalline regions where the reagents concentrate, which finally results in porous hydrogels. The swelling of these "cryogels" attains $40–70 \, ml \cdot g^{-1}$, the average diameter of pores is 40–80 μm [36].

All known methods of initiation are used for inducing three-dimensional polymerization of AAm, AAc, and their copolymerization [37, 38]. Some special initiating systems have been also elaborated [39]. There is a wide choice of crosslinking agents: various *N,N'*-bis-acrylamides (methylene, ethylene, isopropylidene) as well as dimethacrylates of oligoethylene glycols and some others.

Radiation initiation with simultaneous crosslinking of the resulting polymers also seems attractive. In this case purity of the reaction system is not so important. Thus, γ-irradiation of 5–20% AAm solutions in air at doses up to 0.2 Mrad leads to the PAAm gels with a swelling of $50–100 \, ml \cdot g^{-1}$ [40]. A greater degree of swelling is difficult to attain due to contradictory requirements for swelling and the sol fraction. However, the problem is likely to be solved by incorporating ionic groups into these hydrogels. Similar radiation crosslinked PAAm can be obtained directly from the polymer (see Sect. 2.2.2).

Most other hydrophilic monomers inclined to radical polymerization (methacrylamide, methacrylic acid, *N*-vinylpyrrolidone, aminoalkyl and hydroxyalkyl methacrylates) do not form hydrogels with high swellability in water.

2.1.2 Graft Polymerization

A large number of SAHs described in the literature combine synthetic and natural macromolecules in the network structure. The natural components are usually starch, cellulose, and their derivatives. It is assumed that introduction of rigid chains can improve mechanical properties (strength, elasticity) of SAH in the swollen state. Radical graft polymerization is one of the ways to obtain such SAH.

In some cases, the carbohydrate component obviously performs the role of a crosslinker. The grafting of AAm on multifunctional allyl ether of carboxymethyl cellulose (AE CMC) is a recent example [41, 42]; some date on hydrogels of this type are given in Table 2. As can be seen, a monotonous increase in swelling with a decreasing content of AE CMC is evident, the estimates of the network density from the elastic modulus indicating the quantitative participation of allyl groups in the reaction. Ionogenic units of AAc sharply increase the swelling degree, their effect being independent of the introduction mode (copolymerization or hydrolysis), which shows that the electrostatic contribution to swelling does not depend on the type of charge distribution in the network.

A great variety of SAH was obtained using radical grafting of hydrophilic monomers on natural polymers without any previous modification; Table 3 represents a short summary of the data on these SAH which can be divided into two groups.

In one of these groups the polymer network consisting of synthetic and natural polymers is formed directly in the process of grafting. The degree of swelling is determined by the reaction conditітions including the component ratio, initiation method, and ionogenic monomer content.

The second group includes SAHs obtained by radical grafting of acrylonitrile (AN) on natural polymers, mostly starch, under the action of cerium initiators [43–46, 50, 51]. The proper crosslinked hydrophilic polymer is formed at the stage of alkali hydrolysis of grafted polyacrylonitrile (PAN), the final characteristics depending on many factors, in particular the sort of starch [46], the methods of its preparation [51], the component ratio, etc. The nature of starch is exhibited through

Table 2. Swelling of polyacrylamide hydrogels crosslinked with allyl ether of carboxymethyl cellulose (AE CMC)[1] [41]

Content of AE CMC, wt%	Swelling degree, $ml \cdot ml^{-1}$	
	neutral[2]	ionized[3]
0.027	60.7	5040
0.053	27.2	2450
0.117	20.4	1000
0.254	16.2	610
0.667	13.3	230

[1] Polymerization in 30% aqueous solution of AAm.
[2] In 0.1 N aqueous NaCl.
[3] Hydrolysis in 1 N solution NaOH; extent of hydrolysis ca. 40%.

Table 3. Network graft polymers based on synthetic and natural macromolecules

Biopolymer	Grafted monomers	Method of grafting[2]	Swelling in water, ml · g^{-1}	Refs.
Starch	AN	a	400–1700	[43, 44]
Starch	AN	a	300–1550	[45, 46]
Starch	AN, AAm	a	1200	[44]
Starch	AN, AAc	a	1500	[44]
Starch	AAm	b	12	[47]
Starch	AAm, AAm-SO_3H[1]	b	160–540	[47]
Starch	AAc, AAm	a	290–440	[48]
Starch	AAc, AAm	b	500–1000	[48]
Amylose	AN	a	300	[46]
Cellulose	AAc	a	48	[49]
Hemicellulose	AN	a	430–500	[50]

[1] 2-Acrylamido-2-methylpropansulfonic acid.
[2] a – radical initiation, b – γ-irradiation

the ratio between amylose and amylopectin since the grafting on the latter is much more effective [45]. It is necessary to very carefully choose the optimal H-SPAN (hydrolyzed starch-PAN copolymer [44]) synthesis conditions by extensive study.

The availability of the initial material and its relative cheapness make this type of SAH highly attractive from a practical viewpoint. They are manufactured commercially by a number of companies for agricultural, sanitary, and hygienic applications.

Similar hydrogels including cellulose do not show high swelling due to low hydrophilicity and structural peculiarities, thus making additional modification necessary.

2.2 Crosslinking of Water-Soluble Polymers

The majority of hydrophilic and water-soluble polymers are manufactured on the commercial level. Their list includes PAAm, copolymers of AAm, PAAc, PEO, PVA, poly(*N*-vinylpyrrolidone), some polyamines etc. So, it seems practically highly attractive to produce hydrogels based on these polymers using various crosslinking techniques.

2.2.1 Chemical Methods

Crosslinking using reactions of multifunctional reagents with side groups of polymers is most frequently applied to PVA, some biopolymers containing OH-groups, partly hydrolyzed PAAm, polyamines, and other polymers. These reactions are reviewed in Ref. [52].

Aldehydes (glutaric, succinic [53–56]) are used to crosslink PVA, the network density being controlled by the ratio of aldehyde to OH-groups. The swelling degree of these hydrogels does not exceed 35–40 ml · g^{-1}, and there are almost quantitative correlations of the network density with the crosslinker concentration

[55, 56]. The reaction with epichlorohydrin (ECH) results in even more dense PVA-based networks (swelling up to 2.3 $g \cdot g^{-1}$ [57]). The absence of ionic groups in crosslinked PVA does not allow to reach high degrees of swelling, therefore the published data on PVA hydrogels with the swelling of 370–490 $ml \cdot g^{-1}$ [58] appear to relate to copolymers with ionogenic monomers.

The reaction with ECH is widely used for polysaccharide crosslinking to obtain chromatographic gels with a swelling degree of up to 10–15 $ml \cdot g^{-1}$. The swelling of similar purely cellulose hydrogels (3–6 $g \cdot g^{-1}$ [59]) is even lower and, therefore, it is necessary to modify them by destroying the structure or by introducing ionogenic groups. The swelling degree of crosslinked starch can be increased from 50 to 450 $ml \cdot g^{-1}$ by carboxylation with monochloroacetic acid [60].

There is a variety of multifunctional reagents which are used for crosslinking the polymers by their side groups. Recently it has been shown that various polyamines can be effectively crosslinked by reacting with bis-epoxides [61], the resulting network possessing all the features of partly ionic hydrogels with a swelling degree of up to 25 $ml \cdot ml^{-1}$. The reaction with polyisocyanates is another promising methods of chemical crosslinking of polymers containing active hydrogen atoms in their side groups (OH, NH, etc.). This possibility can be illustrated by the synthesis of neutral networks based on poly(*N*-acetylethyleneimine) obtained by polymerization of 2-methyl-2-oxazoline followed by partial hydrolysis [62]. The reaction of the NH-groups of the polymer with hexamethylene diisocyanate results in networks with swelling depending on the hydrolysis degree and the crosslinking agent content (Table 4). Hydrogels with the highest swelling are obtained at nonstoichiometric crosslinking which provides a more loose network and partly retains ionogenic groups. The presence of the electrostatic contribution is confirmed by swelling suppression achieved by adding salt.

Irreversibility and a wide choice of catalysts for regulating gelation are the advantage of the two latter methods of crosslinking, their shortcoming being the impossibility of carrying out the reaction in aqueous medium.

Interaction with plurivalent cations via ligand exchange mechanism is one more rather widely applied crosslinking technique. The network bonds of ionic or donor-acceptor nature are located, with respect to lifetime, between the truly covalent crosslinks and physical entanglements. Generally speaking, gelation in these systems is reversible.

Table 4. Swelling of hydrogels based on partly hydrolyzed polymers of methyl-2-oxazoline [62]

Degree of hydrolysis, %	Molar ratio NCO/NH (at crosslinking)	Swelling, $g \cdot g^{-1}$	
		in water	in 5% NaCl
18.2	1.21	2	2
7.2	0.98	10	8
3.4	1.19	14	11
3.4	0.65	45	19
3.9	0.59	72	25

Another interesting approach to SAH production is the crosslinking of partially hydrolyzed PAAm with salts of multivalent cations, particularly, Cr^{3+} [63–68]. As a rule, a macroscopic gel is formed at concentrations above the coils overlapping point, with intramolecular crosslinking prevailing below this point. The methods of equilibrium dialysis [63], rheology [64–66], spectroscopy [67], and others are found to be convenient for observing the dynamics of gelation in the PAAm-Cr^{3+} system. The possibility of directly calculating the number of cations bound, i.e. the network density, and elastic modulus measurements makes such systems a useful model for studying the gelation process. However, a correlation between these values expected from the gelation theory was not observed [64, 68].

Unfortunately, reliable data on the equilibrium swelling and long-term stability of hydrogels obtained by this method are not available. It should be noted that crosslinking of polymers with metal ions is one of the few reactions which can be carried out directly in the soil medium, which is efficiently used, e.g., in oil recovery [65, 66].

A similar example is the formation of nonstoichiometric interpolymeric complexes between mutually complementary polyelectrolytes – polycation and polyanion [69, 70]. They behave like true polymer networks and are capable of swelling; the interpolymeric complexes between PAAc and polyethylene piperazine swells, for instance, 16–18 times [70]. Also advantageous in this case is the possibility to carry out this type of crosslinking in open systems, such as soil.

Generally speaking, in preparation of weakly crosslinked networks from preformed polymers one faces the same problems as in free- radical polymerization, namely, kinetic restrictions due to low concentration of crosslinker groups, their participation in intramolecular reactions. The gelation process can also come out of control due to a poor compatibility of crosslinking agent with the polymer as well as nonuniformity of its distribution in the bulk of polymer or in viscous solution. For this reason, other crosslinking techniques based on physical effects without introducing any new reagent are preferable.

Gelation induced by thermal treatment is known in the case of PVA; the swelling degree (10–15 $ml \cdot g^{-1}$) and the modulus of elasticity regularly changes with the number of freezing – thawing cycles [71]. Thermal crosslinking is also characteristic of PAAm and hydrolyzed PAN where it proceeds according to the transimidization mechanism. This technique is used to obtain SAH based on hydrolyzed PAN grafted on starch [72].

2.2.2 Radiation Crosslinking

This is one of the most universal techniques for obtaining hydrogels from water-soluble polymers. Crosslinked PEO, PVA, PAAm, PAAc and its salts, as well as some polymer blends were obtained by this method. Although all polymers mentioned above have their own specific features, in most cases the gelation doses do not exceed 1–2 Mrad, i.e. they are substantially lower than for the same polymer in bulk. This is due to the fact that in aqueous media crosslinking occurs indirectly, namely because of the $OH^{\cdot}$ radical formation and their attack on the macromolecules. There exists a developed theory of these processes [73].

Detailed studies on radiation chemistry of PEO have been performed [74–77]. Upon γ-irradiation, the gel-dose drops abruptly along with an increase in the concentration and molecular weight of the polymer, thus reaching values of 0.15–0.25 Mrad in the range of practical interest [75]. Oxygen is a strong inhibitor and when it is carefully removed from the solution, crosslinking of PEO occurs at doses as low as 0.01 Mrad [76].

After crosslinking under γ-irradiation, the structure and morphology of PEO undergo substantial changes: the melting point and the degree of crystallinity decrease with the dose, possibly due to a decrease in the effective chain length and steric hindrance caused by crosslinking. Evaluation of the network density of radiation crosslinked PEO based on elastic modulus gives values from 1.14×10^{-4} to 2.34×10^{-4} mol · cm^{-3} in the dose range of 1–10 Mrad [77]. Radiation crosslinking of PEO in solutions is followed by vigorous macrosyneresis (Table 5). The equilibrium swelling of hydrogels forming after irradiation increases with dilution of the irradiated solution; however, when drying, hydrogels irreversibly lose some swelling liquid which may be a result of a partial crystallization. Similar assumptions were also made with respect to PEO networks obtained by chemical methods. Radiation crosslinked PEO with swelling of about 100 ml · g^{-1} was also obtained in patents (for example, Ref. [79]) and one can suggest that the first experiments with plants [7–9] made use of these particular SAHs.

In recent years, a number of works aimed at obtaining SAH based on PAAm specifically for agricultural applications [13, 80–82]. Some of these SAH are believed to be tested or already used on the commercial scale. Radiation methods are considered to be most rational for producing PAAm-based SAH. Features of the PAAm crosslinking are quite typical for this entire group of polymers. Lower sensitivity to oxygen and the possibility to crosslink slightly wet (up to 10–20% of water) PAAm beads are observed; the gel-doses do not exceed 2 Mrad [81]. Swellability of radiation crosslinked PAAm has not been thoroughly studied yet.

There are several other interesting polymers forming SAH with swellings up to 1500 ml · g^{-1} under irradiation in aqueous solutions, such as: sodium salts of PAAc [83], copolymers of AAm with AAc [22], poly-*N*-vinylpyrrolidone [84], PVA

Table 5. Radiation-induced crosslinking of PEO[1] [78]

Concentration of polymer, g · dl^{-1}	Syneresis, vol. %	Swelling of hydrogels, ml · g^{-1}	
		before drying	after drying
3.0	0.2	35	41
2.0	1.5	50	38
1.33	2.3	74	38
0.82	6.0	115	42
0.57	46.3	95	38
0.38	55.3	123	33
0.38[2]	24.0	189	34

[1] $\bar{M}_v = 5 \times 10^5$, γ-irradiation ^{60}Co, 0.5 Mrad.
[2] 0.2 Mrad

[85], and some others. However, the theory of gelation under radiation has not yet been effectively used to optimize the production and properties of SAH. Exceptions are given in Refs. [84, 86].

Of some interest is also co-crosslinking of various synthetic polymers, their blends with natural ones as well as compositions with inert or active fillers; numerous patents are devoted to these materials (for example, Refs. [87, 88]). Low doses of crosslinking allow to introduce various physiologically active additives into SAH without any danger of radiation damage. This possibility is particularly attractive for the technology of SAH.

2.3 Other Reactions

Among other approaches to the synthesis of SAH are also reactions in the polymer chains and polyaddition.

2.3.1 Polyaddition

Crosslinking of poly(ethylene glycols) (PEG) with triisocyanates [89] or multifunctional isocyanates [90] is the most typical example of applying this reaction to hydrogels. A precursor determines the chain length between crosslinks in the resultant polyurethane network, which allows to predict quantitatively the network density. Another variant is the reaction of PEG with diisocyanate and triol, which is a branching agent [91]. Availability of PEG in a wide range of molecular weights makes these reactions attractive for practical syntheses and a convenient model of gelation.

Table 6. PEO hydrogels obtained by polyaddition

Molecular weigth of PEG	Conc. of PEG, wt%	Method of synthesis [2]	Content of water in swollen SAH, %	Swelling [3], $ml \cdot ml^{-1}$	Refs.
1010	20	a	80.9	5.93	[90]
1810	20	a	83.9	7.16	[90]
3000	[1]	b	84.1	—	[89]
3400	20	a	96.6	34.64	[90]
5650	20	a	96.7	36.36	[90]
5650	33	a	92.3	15.54	[90]
5650	50	a	91.0	13.26	[90]
5650	in bulk	a	—	4.78	[90]
8250	20	a	96.8	37.47	[90]
8310	in bulk	c	—	~ 11.5	[92]
8630	[1]	b	75.2	—	[89]
20000	[1]	b	55.1	—	[89]

[1] From 20 to 50 wt%.
[2] a — pluriisocyanate Desmodur N 75; b — tris(4-isocyanatophenyl)methane; c — methylene bis(4-phenyl-isocyanate) + 1,2,6-hexane triol.
[3] In water at 60 °C

Swelling parameters of some polyurethane hydrogels are given in Table 6. Two distinct trends can be obviously observed in the Table. The first is an increase in swelling with increasing PEG molecular weight, which is in good agreement with the expected tendency. Simple estimates show that swelling of a series of hydrogels obtained in Ref. [90] is in a complete accord with the theory (at $\chi = 0.465$ [92]). The second trend, i.e., the increase of swelling with dilution of the reaction mixture, is not so evident. However, it may be due to intramolecular cyclization and a decrease in the effective number of crosslinks.

As can be seen from Table 6, the results obtained by various groups of authors are markedly different in the region of high molecular weights of PEG. This may be caused by the presence of uncontrolled proton-containing impurities participating in the reaction together with the OH-groups of PEG. The role of these impurities should increase along with a decrease in concentration of the OH-groups in the reaction mixture. Another reason for an ambiguity in the hydrogel properties may be microsyneresis and formation of additional crosslinks due to crystallization of polymer [90, 93].

The PEO-based polyurethane networks can hardly be considered as a practical perspective in agriculture, though in principle this technique for obtaining SAH should not be neglected.

2.3.2 Reactions in Polymer Chains

Reactions of this type are quite popular and widely used to introduce hydrophilic and ionogenic groups into linear polymers as well as directly into polymer networks. These reactions include hydrolysis (PAAm, PAAc and their analogs from PAN, PVA from poly(vinyl acetate), oxyethylation and oxymethylation of starch and cellulose, sulfurization, and other reactions. These processes are of industrial importance, well studied and widely reviewed.

Of prime importance for obtaining SAH are the reactions of PAAm hydrolysis yielding copolymers of AAm and AAc with different proportion of units, structure, and properties [23, 38, 94, 95]. Hydrogel swelling substantially increases when this reaction is performed directly in these gels; for example, alkaline hydrolysis increases the swelling degree up to 12–16 times in the case of crosslinked PAAm [21] and up to 80–90 times in PAAm grafted on cellulose [41]. Under the alkaline hydrolysis conditions, the reaction in restricted in conversion by the theoretical limit but proceeds to a greater extent upon acid catalysis.

It should be noted that hydrolysis of the PAAm chains in hydrogels can also occur in aqueous medium at the application conditions, i.e. in the soil. This should obviously cause an increase in swelling but at the same time it makes the risk of irreversible ionic collapse more real. This tendency is irresistible and can serve as one of the factors determining the lifetime of SAH in soil. Abrupt disappearance of the SAH effects were observed in some of our long-term experiments.

The above reviewed techniques permit to obtain hydrogels with a desirable set of properties, primarily swelling, depending on the choice of the starting material. Some of these methods, namely crosslinking polymerization, grafting on biopolymers, chemical and radiation crosslinking are already a base of hydrogels commercial production. According to available estimates, the total world output of various sorts of hydrogels approaches 100,000 tons a year.

3 Key Physical Properties of Hydrogels

Undoubtedly, the properties of superabsorbent hydrogels occupy the key position in the problem under consideration. Being directly connected with the network formation reaction, they provide all necessary information about the details of this process. Also, the SAH properties are found to be the most reliable basis for understanding and predicting their behavior in real systems, i.e. in the soil, in contact with plants, in physiological media, etc.

The main property of agricultural SAH is their ability to absorb, retain in the swollen state, and then to transfer large volumes of water, in other words, their swelling behavior in a broad sense. In this section, we consider the main features of the behavior including, when necessary, some fragments of the theory of these systems and methods of their structural analysis.

3.1 Properties in Swollen State

The parameters which characterize the thermodynamic equilibrium of the gel, viz. the swelling degree, swelling pressure, as well as other characteristics of the gel like the elastic modulus, can be substantially changed due to changes in external conditions, i.e., temperature, composition of the solution, pressure and some other factors. The changes in the state of the gel which are visually observed as volume changes can be both continuous and discontinuous [96]. In principle, the latter is a transition between the phases of different concentration of the network polymer one of which corresponds to the swollen gel and the other to the collapsed one.

Taking into account the usual conditions of the hydrogel application, we shall limit our discussion to that part of the phase diagram which depicts the swollen gel and touch upon the collapse only in passing since a detailed analysis of this phenomenon is given by Tanaka [96, 97], Ilavsky [19, 98], and Khokhlov [99].

3.1.1 Experimental Methods

Progress in the theory and advances in practical applications of hydrogels are to a great extent determined by experimental study of their swelling and elasticity. However, investigating SAH is rather complicated because most of the available techniques are adapted mainly to highly crosslinked gels.

The traditional methods for measuring the swelling degree [100] are, as a rule, limited due to the difficulties in quantitative separation of the swollen gel from the outer solution because of extremely low strength of the former. These difficulties can be avoided by measuring the dimensions of a regular shape sample directly in an excess of liquid [19, 101, 102]. The other example is the modified volumetric method recently developed by us especially for SAH [103].

Direct mechanical methods can be used to determine the swelling pressure of hydrogels, e.g., by means of devices in the form of a cylindrical chamber equipped with a piston in which the gel contacts the solution through a porous membrane. This technique allows measuring very low pressure (of the order of 0.1–10 kPa) and makes it possible to analyze the SAH with swelling up to 700 ml · g^{-1} [102, 103]. Among others, the method of osmotic deswelling is to be mentioned [104].

The methods for measuring the elastic modulus of gels are reviewed in Ref. [105]. Direct measurements of equilibrium stress-strain isotherms of SAH are complicated by the gel softness. Nevertheless, a number of experiments on compression and tension of the gels has been reported (see, for example, Refs. [18, 21, 42]). The method of dynamic light scattering is free from such inconveniences [106]. However, it yields dynamic modulus. Some other techniques were also used to characterize hydrogels, for example, viscoelastic measurements [28, 30, 31] and swelling equilibrium [20].

3.1.2 Equilibrium Swelling Degree

Of particular importance for the application are the effects of the external compression and the ionic composition of the outer solution on the swelling degree. The reason is that hydrogels usually exist in mineralized aqueous solutions (soil solution) and are affected by compression, for example, produced by the surrounding particles of the soil. Even in the absence of any external load the compression develops due to the gel swelling in a constrained volume.

Now, we consider the situation where the external mechanical forces are absent, that is, the case of "free" swelling.

The effect of the ionic composition of the medium is most markedly pronounced in polyelectrolyte hydrogels. Reversible swelling or contraction in response to changes of pH and the ionic strength are typical of these gels [96]. The equilibrium swelling curves of the SAH in aqueous solutions of low-molecular-weight electrolytes are shown in Fig. 2. The ions present in the solution strongly suppress swelling, and the higher the ion charge, the lower their concentration ensuring the same degree of deswelling. Exceptions include H^+ and other ions giving stable associates with the ionized network-fixed (usually carboxylate) groups. Other conditions being the same, these ions exhibit the strongest suppression effect.

When the ionic composition varies within the range of very low salt concentrations, the swelling degree can reach its maximum [101]. This behavior is observed

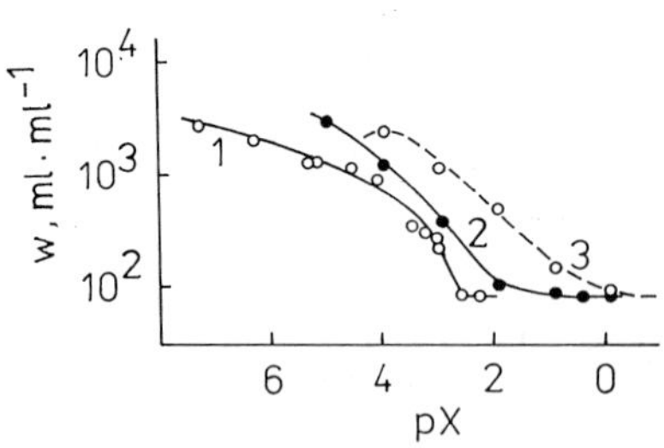

Fig. 2. Dependencies of the swelling degree of PAAm hydrogels on the concentration X (in mol · L^{-1}) of H^+ (*1*), Ca^{2+} (*2*), and Na^+ (*3*) in the solution. *Dashed curve* 3 is obtained by calculation. From Dubrovskii et al. [22]

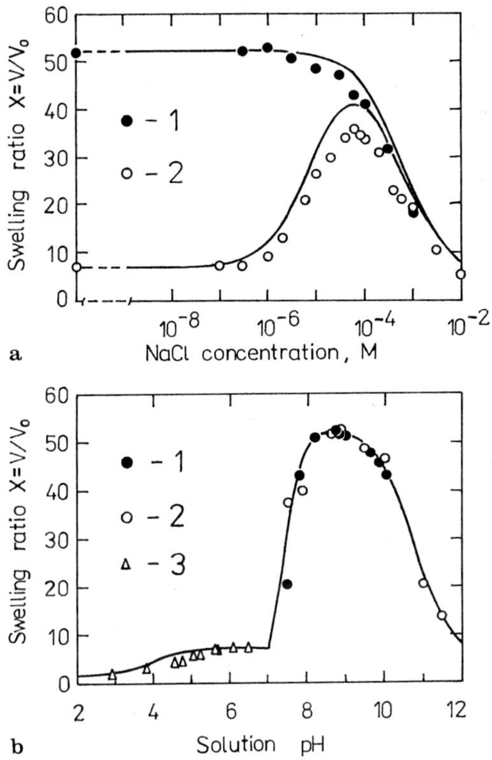

Fig. 3a, b. Dependencies of swelling ratio of PAAm hydrogel (5.2 mol % of AAc units) on concentration of NaCl (**a**) and pH (**b**). *Solid lines:* theoretical predictions. Experimental points: **a** — pH 8.7 (*1*), 6.5 (*2*); **b** — solutions of triethanolamine (*1*), NaOH (*2*), acetic acid (*3*). From Rička and Tanaka [101]

in neutral solutions (Fig. 3a) as well as in basic solutions of varying pH (Fig. 3b). It should be noted that swelling maxima in salt solutions are observed only when deionized water is used as solvent. Within the practically important range of external conditions the swelling degree decreases monotonously with increasing concentration of ions.

All the features considered above can be easily interpreted in terms of the somewhat modified theory of swelling [4] which satisfactorily describes the behavior of weakly charged gels [101, 102]. In particular, according to the theory, the increasing concentration of ions forces the swelling degree towards the lower limit (see Fig. 2) which corresponds to the swelling of the uncharged network (w_{net}) and is determined only by the crosslinking density and the interaction parameter.

In most cases, the swelling of polyelectrolyte hydrogels depends only on ionic strength of the solution but not on the size and nature of the ions [101]. Therefore, the ionic suppression curves similar to those of Fig. 2 and 3 are to some extent universal and allow to predict quantitatively the swelling of hydrogels for practically any ionic situation.

The swelling behavior of hydrogels in solutions of multivalent ions capable to associate with the network-fixed charges, e.g., Cu^{2+}, substantially differs from that described above, viz. the collapse of gels takes place [107]. As a result of this

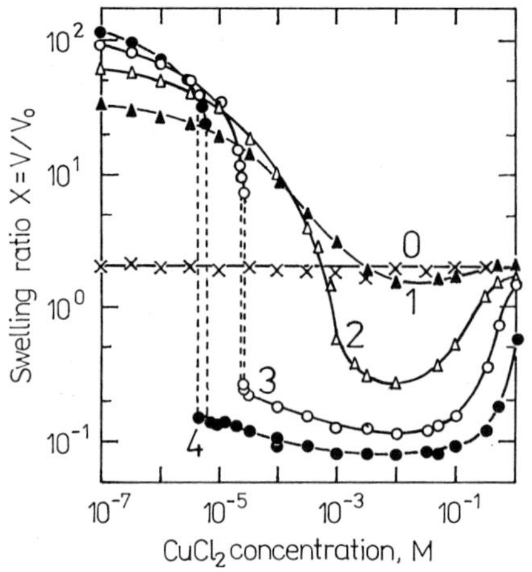

Fig. 4. Swelling ratio of PAAm hydrogels containing varying amount of AAc units as a function of $CuCl_2$ concentration in the outer solution. Curves are numbered with respect to increasing ionic group content: 0 (*0*), 36.5 (*1*), 107 (*2*), 145 (*3*), and 212 mM (*4*). From Rička and Tanaka [101]

transition, which is attributed to the formation of energetically favorable contacts between polymer segments due to the association of a copper ion with several network-fixed ligands, the usual dependence of the swelling degree on the ionogenic group content of the gel is reversed and the swelling degree of the charged gel proves to be much lower than the w_{net} value (Fig. 4). Similar dramatic consequences can apparently be caused by some other multivalent cations (Mg^{2+}, Ca^{2+}, Al^{3+}, Fe^{3+}, Cr^{3+}) [108], which should be taken into account in designing hydrogels for practical purposes.

In contrast to polyelectrolyte hydrogels, nonionic ones are almost insensitive to the ionic composition of the medium. In particular, the equilibrium swelling degree of a charge-free PAAm gel practically does not depend on the Cu^{2+} concentration (Fig. 4, curve 0) nor on the concentration of several other ions or on the pH value [101].

In the case of PEO gels, the swelling degree shows also a negligible dependence on pH of the solution [109]. At the same time, taking into account the behavior of the linear PEO in solutions [110], one can expect that the swelling degree of PEO gel will decrease with an increase in the ionic strength. The fact is that the linear PEO is salted out in aqueous solutions, as salts reduce solution viscosity. This salting-out effect, though mild in comparison with that observed in polyelectrolytes, should manifest itself also in swelling of PEO networks.

These examples show once again that the network-fixed charges play an important role in the swelling behavior of hydrogels.

3.1.3 Swelling Pressure

When considering the effect of the external compression on the SAH swelling their response to mechanical forces is of primary interest. Such forces arise, for instance, when the gel swells in the cavity constraining its volume. In this case the gel swelling pressure (π) expanding the network is counteracted by an external pressure (p) due to the reaction at the cavity walls. The osmotic compression can produce the same effect. In both cases the external pressure growth results in a decrease in the equilibrium volume and an increase in the swelling pressure of the gel according to:

$$\pi(w) = p \tag{3.1}$$

The only significant difference between the osmotic and mechanical forces is that the latter may not be isotropic as, for example, at the uniaxial swelling of the gel in a tube.

Figure 5 shows the results of experiments on SAH swelling in a cylindrical cell permeable for the solvent [22]. The data exhibit a strong dependence of the swelling degree on the small load applied to the gel by the piston. An ambiguity of the dependencies w(p), particularly noticed at very small pressures, is an important feature of SAH. Other conditions being the same, the equilibrium volume of SAH depends on whether the swelling proceeds at a fixed volume (curve 2) or at a fixed load (solid curve 1). In the first case, adequate to free swelling, the network deformation proceeds isotropically, in the second case – uniaxially; at uniaxial swelling the sample is compressed in the two other directions and therefore its equilibrium volume is lower than at free (isotropic) swelling [102, 111]. This effect is especially evident at great changes of the gel size, that is namely in the SAH case.

Osmotic deswelling experiments, performed with a series of PAAm [20] and PVA [112] gels have revealed a correlation between their sensitivity to the external pressure and the equilibrium swelling degree in the absence of external forces. This fact is illustrated below:

	PAAm [20]					PVA [112]			
w, $ml \cdot ml^{-1}$	16.0	20.3	27.1	43.3	144	9.6	16.6	27.8	58.8
Δw*, $ml \cdot ml^{-1}$	3.0	5.0	10.0	23.5	122	1.1	4.4	13.1	39.6

* at 10 kPa

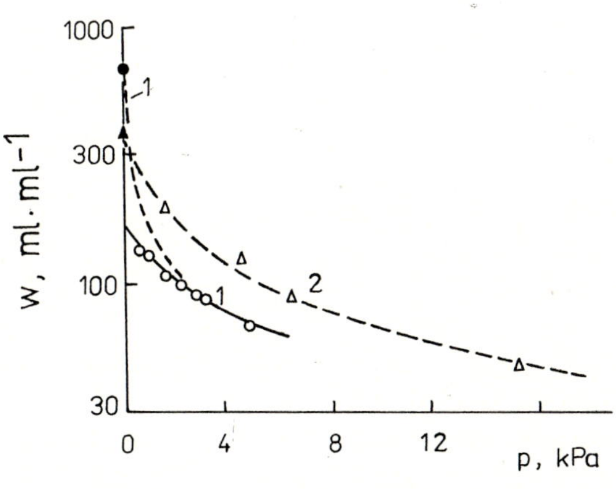

Fig. 5. Influence of pressure on swelling of two samples of partly ionized PAAm hydrogels in 0.01 N aqueous NaCl solution; solid curve – uniaxial, dashed curves – isotropic swelling as obtained by calculation. Points on the y-axis correspond to the free swelling. From Dubrovskii et al. [22]

Here the decrease in the swelling degree ($\Delta w = w - w(p)$) due to the same external pressure (the diluent activity) is used to characterize the gel compressibility. The equilibrium osmotic modulus $K_{os} = -w(\partial\pi/\partial w)$ which characterizes the compressibility at zero swelling pressure follows the $w^{-2.23}$ power law for the PAAm gels [20]. All these results show that it is impossible to simultaneously ensure both high swelling degree and low sensitivity of the hydrogel to the external pressure at least by varying the crosslinking density, which also agrees with the theory.

The swelling pressure of polyelectrolyte gels is usually considered as a sum of the network (π_{net}) and ionic contributions (π_{ion}) [4, 99, 101, 113, 114]. The former describes the uncharged gel while taking into account the interaction between the polymer segments and the solvent as well as the network elasticity [4]:

$$\pi_{net} = (RT/v_1)\{(1/2 - \chi)\,\varphi^2 - (\varphi/\bar{x}_c)\,[(\varphi_0/\varphi)^{2/3} - 2/f]\}\,, \tag{3.2}$$

where v_1 is the molar volume of the solvent, $\varphi \ll 1$ is the volume fraction of polymer in the swollen gel, $\bar{x}_c$ is the average number of units in the subchain, $\varphi_0 \simeq \bar{x}_c^{-1/2}$ is the volume fraction of polymer in a reference state [111], f is the crosslink functionality, χ is the parameter of polymer-solvent interaction. For weakly charged flexible networks, the ionic pressure is determined mainly by the difference between the concentrations of freely mobile ions in the gel (c_i^*) and in the outer solution (c_i), i.e. by the Donnan effect [4, 101]:

$$\Delta\pi_{ion} = RT\,\Sigma(c_i^* - c_i) \tag{3.3}$$

The $\pi(w)$ curves calculated from these relations [22, 102] fairly well describe the data on the superabsorbent PAAm hydrogels (Fig. 5). The electrostatic interactions seem to be necessarily taken into account in strongly charged hydrogels [99, 113, 114]. Anyway, the dependence $\pi(w)$ and, consequently, the effect of external pressure can be evaluated theoretically for a wide range of external conditions. The estimates show unambiguously that there exists a direct correlation between the equilibrium swelling of the homogeneous gels and their compressibility.

The swelling pressure or osmotic deswelling data can be, therefore, described as the functions of $\pi(w)$ by either of the theories [115]. This description can be then applied to determining the network parameters (see, for example, Ref. [22]). On the other hand, the swelling pressure which is directly connected with the chemical potential of water in the gel:

$$\Delta\mu_1 = -\pi v_1 \tag{3.4}$$

is a source of information about its energetic state in the swollen network.

3.1.4 Elastic Modulus

Elasticity is another important property of SAH which distinguishes the swollen gel from a viscous (but capable of flowing) solution of the non-crosslinked macromolecules.

An increase in the swelling degree usually results in lowering elastic modulus. According to the rubber elasticity theory [116–118] the shear modulus of the gel G can be expressed as:

$$G = RT\varphi_0^{2/3}\varphi^{1/3}n_c\,, \tag{3.5}$$

where $n_c = 1/(\bar{x}_c v_1)$ is the crosslinking density, i.e. the concentration of elastically active chains in the dry network. Various concentration dependencies of the elastic modulus are observed when the crosslinking density varies and in the case when it remains constant.

The effect of the crosslinking density on the elastic modulus is evidently demonstrated by the data on PEO gels obtained by end-linking of the polymer precursors of various molecular weights [90]. The analysis of these data shows that the dependence of the modulus on the equilibrium volume fraction of the polymer in the first approximation agrees with the scaling predictions [119]. Similar results were obtained for neutral PAAm gels in equilibrium with water [18, 20, 24]. The elastic modulus was found to vary as φ^m with exponent m ranging from 1.88 to 2.55 in accordance with the value predicted for a good solvent. An appropriate example is given in Fig. 6.

The effect of the network density on the polyelectrolyte hydrogel elasticity can be understood taking into account the fact that the elastic modulus is closely connected with the swelling pressure (see, for example, Refs. [20, 115]):

$$G = (RT/v_1)\,(1/2 - \chi)\,\varphi^2 + \Delta\pi_{ion} \tag{3.6}$$

This equation results from Eqs. (3.2) and (3.5) at $\varphi \ll 1$ and corresponds to the gels swollen at equilibrium. Equation (3.6) predicts that at a high ionic strength of the outer solution when $\Delta\pi_{ion} \sim \varphi^2$ [4], the elastic modulus should change as φ^2. At low ionic strengths, the ionic pressure prevails and changes as φ [4]. The elastic modulus is likely to follow the same law and this assumption, according to our analysis, agrees well with the results of an experimental study on PAAm gels containing small amounts of sodium metacrylate [120, 121].

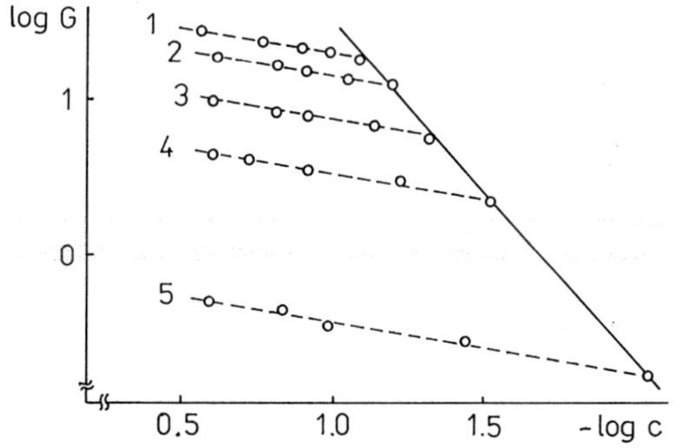

Fig. 6. Double-logarithmic plot of shear modulus G (kPa) vs concentration of PAAm gel C ($g \cdot ml^{-1}$) as a function of deswelling. Samples of hydrogel with equilibrium swelling of 12.3 (1), 15.6 (2), 20.8 (3), 33.4 (4), and 111.1 ml g^{-1} (5). From Geissler et al. [20]

When the gel swelling degree changes at a fixed crosslinking density, then, according to Eq. (3.5), the product $Gw^{1/3}$ should remain constant. For hydrogels this prediction is usually valid. In particular, for the PAAm–PAAc gels with low network density the $Gw^{1/3}$ value is practically independent of the swelling degree varying in a wide range due to changes in the content of network ionic groups [42] or the concentration of low molecular weight electrolyte [22].

Figure 6 shows the shear modulus values for a series of neutral PAAm gels at different stages of deswelling [20]. The slopes of the dotted lines describing the deswelling of each sample are about 0.334, which perfectly agrees with the theory.

By contrast, the concentration dependence of the modulus for the PAAm-based ionic gels with high network density was found to substantially deviate from the law predicted. The product $Gw^{1/3}$ increases considerably with the growth of the swelling degree due to the polymer hydrolysis or when water is evaporated from the gel swollen at equilibrium. This behavior is attributed to the finite extensibility of the network chains which are believed to cease following the Gaussian statistics at high swelling degrees.

Another possible reason for the deviations observed is some heterogeneity of highly crosslinked PAAm gels which manifests itself in particular by an anomalous growth of the modulus at substantial elongations of the gel [18].

The elastic modulus of SAH swollen at equilibrium is rather low. Typical values of the shear modulus and the corresponding swelling degree for some systems under consideration are given in Table 7. As can be seen, the hydrogel elasticity can be controlled within a fairly wide range. However, it is difficult to ensure high levels of both swelling and elasticity, which follows from the above relationships. Heterogeneous networks reinforced by introducing rigid chains (polymers grafted

Table 7. Elastic properties of hydrogels

Type of Hydrogel	Swelling (w), $ml \cdot ml^{-1}$	Elastic modulus (G), kPa	Refs.
PAAm	26	0.05–4.0[3]	[106]
PAAm	6.3–81	0.60–170	[18]
PAAm	16–140	0.16–19	[20]
PAAm–PAAc	13–140[3]	36–69[3]	[21]
PAAm–PAAc	21–2500	2.0–12	[42]
PAAm–PAAc	37–800	3.3–6.2	[22]
PAAm–PAAcNa[1]	48–370	1.7–2.0	[19]
PAAm–PAAcNa[1]	45–440	0.55–17	[120]
PAAm–PAAcNa[1]	21–170	2.1–20	[121]
PEO	3.9–37	2.0–250	[90]
PVA	5.5–26[4]	1.9–67	[53]
PVA	6.7[4]	15–58[3]	[122]
PVP–PMMA[2]	4.8[3,4]	120[3]	[123]

[1] Sodium salt of PAAc.
[2] Copolymer of N-vinylpyrrolidone and methyl methacrylate.
[3] Data obtained from figures in corresponding references.
[4] $ml \cdot g^{-1}$

on starch) or by crystallization can show certain advantages. However, in any case this cannot possibly be done without some loss in swelling, i.e. we are facing a new compromise.

3.2 Network Parameters of Supergels

Knowledge of the network parameters is important for understanding gelation processes, and relationships between the molecular structure and hydrogel synthesis conditions. The principles for the optimization of SAH characteristics for various application purposes can also be based on these parameters.

The SAH network parameters can be determined from the elastic modulus and the equilibrium swelling; however, there are only a few examples of this approach.

To determine the crosslinking density from the equilibrium elastic modulus, Eq. (3.5) or some of its modifications are used. For example, this analysis has been performed for the PAAm-based hydrogels, both neutral [18] and polyelectrolyte [19, 22, 42, 120, 121]. For gels obtained by free-radical copolymerization, the network densities determined experimentally have been correlated with values calculated from the initial concentration of crosslinker. Figure 1 shows that the experimental molecular weight between crosslinks considerably exceeds the expected value in a wide range of monomer and crosslinker concentrations. These results as well as other data [19, 22, 42] point to various imperfections of the PAAm network structure.

Analysis of data pertaining to the modulus of PEO gels obtained by the polyaddition reaction [90] shows that even in this simplified case the network structure substantially deviates from the ideal one. For all samples studied, the molecular weight between crosslinks (M_c^{exp}) exceeds the molecular weight of the precursor ($\bar{M}_n$). With decreasing precursor concentration the $M_c^{exp}/\bar{M}_n$ ratio increases. Thus, at $\bar{M}_n = 5650$ a decrease in precursor concentration from 50 to 20% increases the ratio from 2.3 to 12 most probably due to intramolecular cycle formation.

The swelling equilibrium provides additional information on the hydrogel structure. Interpretation of data on free swelling, swelling pressure, or osmotic deswelling [20, 115] makes it possible to evaluate, besides the crosslinking density, the polymer-solvent interaction parameter χ, the fraction of ionogenic groups β, and their dissociation constant K_H [22, 124].

Various modifications of the Flory theory [4] are usually applied to describe the uncharged gels. Their crosslinking density can be simply calculated from the swelling degree using Eqs. (3.1) and (3.2) or analytical expressions for the M_c value (see, for example, Ref. [124]).

Typical SAH are weakly charged flexible networks. A simple theory was proposed to describe these networks, in which the ionic swelling pressure includes only an ideal (Donnan) contribution [4, 101]. In this approximation, the swelling equilibrium is described by a system of equations including Eq. (3.1) and the following:

$$c_i^*/c_i = K_D^{z_i}, \tag{3.7}$$

$$K_H = \alpha c_H^*/(1 - \alpha), \tag{3.8}$$

$$\Sigma z_i c_i^* - \alpha\beta\varphi/v_1 = 0. \tag{3.9}$$

Equation (3.7) describes the equality of the chemical potentials of the mobile ions on both sides of the gel boundary expressed through the Donnan ratio K_D and the ion charges z_i; Eq. (3.8) concerns the dissociation equilibrium of ionizable (carboxyl) groups of the network; α is the degree of dissociation, c_H^* is the concentration of the hydrogen ions in the gel; Eq. (3.9) represents the gel electroneutrality condition.

These equations allow either to predict the swelling degree ($w = 1/\varphi$) as a function of external conditions or to calculate the network parameters from the correlation between the theoretical and experimental dependencies $w(c_i)$ or $w(p)$ [22, 102]. An example of such a correlation is given in Fig. 2 and 5. As can be seen, theoretical predictions are in good agreement with experimental data. However, when the outer solution contains multivalent cations, only a semi-quantitative agreement is attained.

Complicated theories of ionic gel swelling [99, 113, 114] must inevitably take into account the real electrostatic interactions, the finite extensibility of chains, as well as the electrostatic persistence length effect. Their application is most advisable in the case of strongly charged hydrogels [114].

Returning to the evaluation of the SAH network parameters, it should be noted that the crosslinking densities obtained from the modulus and swelling data agree satisfactorily with each other [22]. Analysis of the data from Refs. [18, 90] confirms this conclusion.

3.3 Kinetics of Swelling

The swelling pattern considered above allows us to understand the peculiarities of the behavior of SAH and the effects encountered during their application. The kinetic aspects of swelling seem to be as important as the thermodynamics of this process. Therefore, we shall touch upon some problems concerning the kinetics of hydrogel swelling and deswelling.

As it has been shown [125, 126], swelling proceeds via the collective diffusion of the network segments in the solvent. In the first approximation, the characteristic time of swelling (and deswelling) of spherical particles is changed proportionally to the square of the gel particle size: $\tau = R_\infty^2/(\pi^2 D)$, where R_∞ is the radius of the gel particle at equilibrium swelling, and D is the collective diffusion coefficient determined by the ratio of the longitudinal network modulus to the frictional coefficient between the network and the solvent.

For small volume changes, the diffusion coefficient is constant and the D values determined by macroscopic swelling experiments are in a remarkable agreement with the values obtained from the gel density fluctuations dynamics. Measurements of the particle sizes [125] and the swelling pressure [103] of the PAAm-type hydrogels during swelling result in D values of ca. 3×10^{-7} cm^2 s^{-1}.

In the case of large volume changes, the situation is somewhat more complicated since the collective diffusion coefficient depends on the concentration and the ionization degree of the gel [127] which undergo substantial changes. Nevertheless, it is possible to find some effective value of diffusion coefficient D from swelling experiments. Thus, according to Ref. [126], this value is $(1-3) \times 10^{-6}$ $cm^2 \cdot s^{-1}$ for isopropylacrylamide hydrogels. The D values of ca. 1×10^{-5} $cm^2 \cdot s^{-1}$, and close to the water selfdiffusion coefficient, were obtained for the PAAm-based SAH using two independent methods, viz. by measuring the gel particle size [103] and the wavelength of the transient pattern on the gel surface during swelling [128].

Therefore, the SAH swelling and deswelling rates can be quantitatively characterized by the time τ which for a given hydrogel type is determined mainly by the gel particle size. The gel instability, both mechanical and thermodynamical, constitutes an additional complication [128–130].

4 Behavior and Effects of Hydrogels in the Soil Medium

The use of SAH in agriculture technologies and plant growing also sets up a number of requirements to be met by these new materials. It is unacceptable to choose the SAH for these purposes just by chance, because it may result in serious economic losses or even in discrediting the idea itself. Therefore, the concluding section of the review is devoted to the peculiarities of the physical behavior of hydrogels in the soil medium as well as to the ideology of searching the proper conditions to realize their maximum efficiency.

4.1 Hydrogels in the Soil-Water-Plants System

The role of water in the life of plants is well known. In terms of its major effects this role consists in transporting the mineral nutrition, maintenance of intracellular pressure responsible for the vertical growth of plants and, finally, participation in photosynthesis which provide the biomass growth, or plainly speaking, the crop production.

Normally, the plants can efficiently use only a small part of the soil water, i.e., from 0.3 to 1 kg of water is required to create 1 g of biomass by means of photosynthesis. The bulk of the soil water disappears through nonproductive channels, which is most typical of sandy soils.

Figure 7 shows schematically the main components of the water balance in soils, the integral equation for which in terms used in the figure can be written as follows:

$$W + Tr = (Pr + Ir) - (Ev + Fl) \tag{4.1}$$

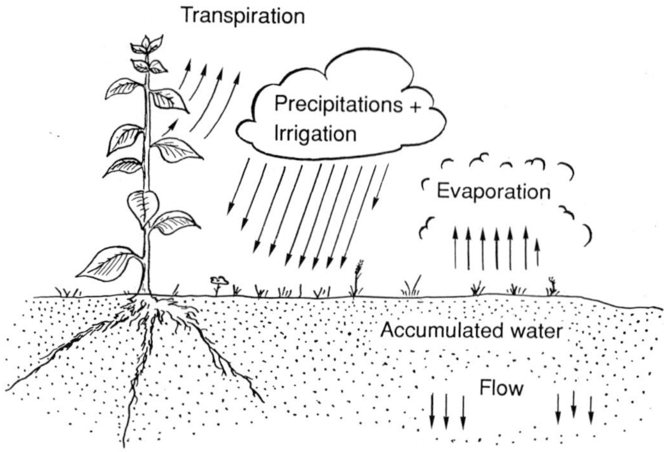

Fig. 7. Main contributions to the water balance in soil-plant system

The left-hand part represents the useful fraction of water which is the difference between its input in the form of precipitations (Pr) or irrigation (Ir) and losses via evaporation (Ev) and gravitational flow (Fl). In terms of Eq. (4.1), the role of hydrogels is mainly determined by their ability to increase the water-holding capacity of the soil (W) by decreasing all the losses.

Experiments made by a group of soil scientists from Egypt [10–13, 131] have explained the physical mechanism of hydrogel action on soil properties; some data from these works are summarized in Table 8. As can be seen in the Table, the hydrogel particles placed in the pores of a sandy soil enable the large capillaries to hold water, the same capillaries in the control soil do not exhibit such ability. This tendency is manifested in an increasing amount of small capillaries and the diminishing average diameter of the pores.

Another group of effects consists in blocking the channels of losing water from the soil layer, i.e., the hydraulic conductivity responsible for the gravitational flow, and of physical evaporation. All these effects provide an increase of the water content of the soil and, consequently, improve the water supply of plants, which is reflected in the three last columns in Table 8. According to the data of various authors, an increase in the soil water content (ΔW) in sandy soils lies in the range of 10–35% at doses up to 0.2%; in a number of cases [10, 11, 58, 131–133] the dependencies of ΔW on the doses of the hydrogels added have been studied.

In first approximation, this dependence can be presented as a linear one until a free volume of the soil pores is attained, i.e., up to 35–45%. Consequently, an instant increase in the soil water content can be written as:

$$\Delta W = wD\,, \tag{4.2}$$

where D is the dose (%), w is the swelling degree ($g \cdot g^{-1}$), which allows to make some estimates. Thus SAH with a swelling of 200 and 500 $g \cdot g^{-1}$ will give a 10%

Table 8. Effect of supergel on soil properties and plants growth [10, 131]

Concentration of gel[1], %	Soil porosity, %			Hydraulic conductivity, $m \cdot day^{-1}$	Avialable moisture[1], %	Rate of evaporation[2], arb. units	Transpiration ratio[3], $g \cdot g^{-1}$	Water use efficiency[4], $g \cdot kg^{-1}$
	Total	Micro/total	Mean pore diameter, μm					
Control	41.32	11.67	19.29	9.75	1.46	92.0	1209.83	0.8265
0.05	42.90	13.05	18.53	9.00	6.71	91.0	1088.12	0.9190
0.10	45.66	15.27	16.63	7.25	7.72	90.0	1057.63	0.9455
0.15	46.51	17.82	15.63	6.32	8.94	89.0	889.81	1.1238
0.20	47.83	30.98	12.34	3.99	11.69	87.0	937.39	1.0668

[1] On dry weight basis.
[2] Adjusted water loss via evaporation.
[3] Grams of evapotranspired water to produce 1 g of dry matter.
[4] Grams of dry matter produced per 1 kg of water.

moisture increase (or 50 mm of water per 35-cm soil layer) at the doses of about 3 and 1 tons ha^{-1}, respectively [48]. All this gives only a notion of the scale, being rather far away from reality.

For some reasons it is practically impossible for a hydrogel introduced into the soil to attain its free swelling and, therefore, the dependencies W(D) differ from those calculated and, as a rule, are nonlinear. It is true, however, that some rare exception exist [58]. We shall come back to this problem below.

If we make an attempt to express formally a complete biological efficiency (E) of SAH as a soil additive, we should additionally consider the multiplicity of hydrogel swelling (the number of cycle N), as well as coefficients taking into account the extent of realization of swelling in the soil (γ) and the biological availability of water (B), both being smaller then unity. Thus, for E as a biological crop increase per surface unit for the whole SAH lifetime in the soil we can write:

$$E = FB\gamma\omega DN, \quad (4.3)$$

where F is the coefficient of the water used for photosynthesis; E and D can be expressed, for example, in $g \cdot m^{-2}$.

Equation (4.3) is convenient only for classification purposes and does not pretend to be strictly descriptive. Separate contributions into SAH efficiency are considered below.

4.2 Thermodynamic Potential of Water in Hydrogels

The SAH water potential determines many aspects of their behavior in the soil. The processes of water redistribution in the soil, its transport to the plant roots, and assimilation follow the osmotic laws and are regulated by the thermodynamic potential.

In terms of energetics, soil water is classified into several categories [134]: gravitational ($pF < 2$), hygroscopic ($pF > 4.2$ 4.5), and capillary ($2 < pF < 4.2$), where the potential is expressed as $pF = \lg(\pi, \text{cm } H_2O)$. The curves of the water-holding capacity as functions of pF(w) which are the basic hydrophysical characteristic of certain soils determine the relationship between these main kinds of water; various types of these curves are available from the literature [135].

Also the plants are able to use only the portion of water which is in the pF range below 4.2–4.5. This limiting value of the potential coincides with the so-called wilting point, for the osmotic apparatus of most cultivated plants is incapable to suck more strongly bound water from the soil.

It is clear that the potential of water stored in SAH should be strongly correlated with these circumstances, i.e., it should possibly be within the range of 2–4.2. The measurements of the swelling pressure provide the most direct information on the energetic state of water in hydrogels. Figure 8 shows the curves of the swelling pressure and pF depending on the hydrogel swelling degree [48, 102]. As can be seen, only a minute portion of water ($w \sim 2$–4 $ml \cdot g^{-1}$) is not in the pF region available for plants. From the biological point of view it means that much of the

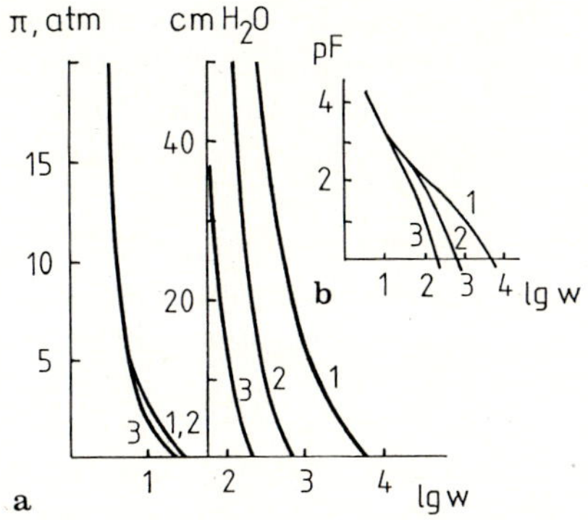

Fig. 8 a, b. Swelling pressure (**a**) and water potential (**b**) of PAAm hydrogel plotted as a function of swelling degree in 10^{-4} (*1*), 10^{-2} (*2*), and 10^{-1} M aqueous NaCl (*3*). Parameters used in calculation: $\bar{x}_c = 4000$, $\beta = 0.2$, pH = 6.3, $\chi = 0.48$. From Dubrovskii et al. [102]

water stored in SAH can be effectively utilized by plant. This finding is confirmed by a number of direct measurements made using soil science methods, including wilting point determinations [10, 48, 58, 131, 136, 137].

Water retained in SAH, however, proves to be considerably affected by the action of relatively weak gravitational potentials. As shown in Figs. 5 and 8, this is particularly the case with contribution to swelling obtained thanks to ionic interactions. Anyhow, under the action of a 10–20-cm water column, the swollen SAH can lose 50–70% of its water. Unlike liquid water, the swollen SAH loses its water according to the deswelling mechanism, i.e., at substantially longer relaxation times. The available data [48, 58, 133] show that the lifetimes of the water in sandy soils with SAH can vary from tens of hours to a few days. The results of water content measurements made in the sand from the Lower Volga region amended with various hydrogels (0.09%) are given below [133]:

	Water content (%) after		
	5 h	46 h	116 h
Control	16.7	12.6	6.1
Acrylate SAH grafted on starch	28.1	23.3	16.5
Crosslinked PAAm	37.8	32.7	25.5

Finally, one more phenomenon is connected with the dependence of swelling on pressure. It is quite clear that the soil layer exerts certain pressure on the SAH particles and this pressure depends on the depth. This factor additively reduces SAH efficiency as a water absorbent. We possess some data, yet unpublished, which directly indicate some decrease of the SAH swelling in soils with the depth of their residence.

There is an opinion in literature that SAH are able to form an additional volume of pores, that is, to provide expansion of soil. Analysis of the π(w) curves shows that this function can be performed only by hydrogels with very low swelling (w ~ 50 $ml \cdot g^{-1}$), so it is doubtful that they can simultaneously ensure a substantial increase in water-holding capacity. These two functions of SAH in the soil seem to be incompatible in terms of thermodynamics, though experiments have indicated an increase in soil porosity [131].

Returning to Eq. (4.3), it becomes clear that low **B** and γ values, particularly the latter, lower the ability of SAH to fulfill their role of water absorbents. In other words, the expected effects can be attained only if a greater dose of hydrogel is introduced, which is not feasible.

Biological effects and stimulation of plant growth induced by hydrogel additives are observed at doses which are often much lower than those obtained from purely physical evaluation. For example, it has been recently shown [13] that, according to various criteria of plant development, the SAH additives even at dosages of 50 to 140 $kg \cdot ha^{-1}$ provide a productivity in sandy soils at the level obtainable by treatment with 20% (of the order of hundreds of tons per 1 ha) alluvial deposits. There seems to exist a mechanism allowing the plants to efficiently utilize small water reserves contained in the SAH particles.

In connection with the thermodynamic state of water in SAH, it is appropriate to consider one more question, i.e., their ability to accumulate water vapor contained in the atmosphere and in the space of soil pores. It is clear that this possibility is determined by the chemical potential balance of water in the gel and in the gaseous phase. In particular, in the case of saturated water vapor, the equilibrium swelling degree of SAH in contact with vapor should be the same as that of the gel immersed in water. However, even at a relative humidity of 99%, which corresponds to pF 4.13, SAH practically do not swell (w ~ 3–3.5 $g \cdot g^{-1}$). In any case, the absorbed water will be unavailable for plants. Therefore, the only real possibility for SAH to absorb water is its preliminary condensation which can be attained through the presence of temperature gradients.

4.3 The Effect of Dissolved Salts

The liquid in which the SAH swelling takes place in real soil (the soil solution) always contains a more-or-less wide set of dissolved salts. Their nature and amount depend on the soil composition, the degree of its salinity, the nature of water entering the soil (rainfall, irrigation, river, or groundwater), the fertilizers used. As a rule, alkali cations, Ca^{2+}, Mg^{2+}, Fe^{3+}, Al^{3+}, and anions Cl^-, CO_3^{2-}, SO_4^{2-}, etc. are the main components of the soil solution; there exist various models of soil solution and nutrient mixtures employed in research, including SAH testing.

As is known (see Sect. 3.1.1), dissolved electrolytes exert a strong depressing effect on swelling of typical SAH, practically eliminating the significant contribution to swelling due to electrostatic interactions. These problems have been broadly discussed [22, 48, 58, 101, 102, 108]. Generally speaking, the benefit obtained owing to the polyelectrolyte swelling proves to be most vulnerable even at low pressures no matter whether this pressure is mechanical or ionic. Nevertheless, the analysis shows that this benefit should not be ignored and polyelectrolyte effects give a rather tangible, though far from maximum, contribution to SAH swelling even in mineralized media.

Depression of swelling by dissolved salts is of serious practical interest since this circumstance results in increasing the SAH dosage, which is economically unprofitable. In his very important work, Johnson [58] gives a detailed description of some typical SAH swelling in aqueous media of various composition; some data from this work are given in Table 9. As can be seen, dissolved salts considerably level off the differences in the „free" swelling of SAH. In extreme situations the water absorption by hydrogels appears to be so minute that it is senseless to use them as soil additives.

The failure in employing PAAm hydrogel (free swelling 1250 $ml \cdot g^{-1}$) in the sandy soil of the Kara Kum desert is due to the same reasons [132]: swelling in rainwater and in saline water from the Large Kara Kum canal used for irrigation purposes was equal to 680 and 255 $ml \cdot g^{-1}$ in the free state and respectively 112 and 76 $ml \cdot g^{-1}$ in the sand. This situation proved to be unacceptable for any SAH application.

Besides nonspecific effects, which can be explained in terms of the ionic strength, such specific phenomena as collapse or crosslinking under the action of multi harged ions are sharply pronounced. They are characterized by threshold phenomena both in the ionogenic group content and the concentration of ions

Table 9. Absorption of water and aqueous solutions by several supergels [58]

Type of water or solution	Starch copolymers		PVA	PAAm
Deionized	1040	1560	488	410
Soft, potable	525	790	216	199
Hard, potable	262	393	150	162
Natural waters				
Upland stream	588	717	237	251
Lowland stream [1]	237	347	131	212
River water [2]	171	148	108	156
Upper estuary	39	42	77	97
Soil-water extracts				
Coastal dune sand	83	108	177	197
Brown earth	237	162	187	212
Gault clay	170	200	167	184
Iron-humus podzol	144	119	139	180
Solutions				
Sodium chloride (45 mM)	92	107	93	125
Calcium chloride (11 mM)	59	63	67	61
Magnesium chloride (13 mM)	73	90	57	70
Magnesium sulfate (10 mM)	67	77	44	64

[1] Receiving run-off from intensively managed agricultural land.
[2] In a major urban-industrial conurbation

responsible for the collapse. Thus, for example, the discrete collapse of the PAAm–PAAc-type gels affected by Cu^{2+} is observed at the ionic content in the network above 0.15, the Cu^{2+} threshold concentration decreasing with growing β value, and at $\beta \sim 0.31$ it is the equal to 5×10^{-6} M (see Fig. 4). Collapse of these gels in the presence of Ca^{2+} occurs at cation concentrations above 10^{-3} M and ionogenic group content above 50% [108].

The salt attack is also an important factor determining the SAH efficiency in the soil medium. In terms of Eq. (4.3), it is manifested by a sharp decrease of the coefficients γ and B. The hydrogel structure prediction for specific application conditions requires to take into account universal (ionic strength) and specific (collapse) suppression phenomena and, therefore, a rather delicate balancing in search for a compromise between swelling gains due to the network density (n_c) and the ionicity (β).

The "golden mean" can be found with the help of laboratory tests used in soil science or by calculation. Purposefully, for this particular review we performed a computer modeling in order to differentiate between various contributions to swelling as well as to estimate γ and B coefficients in Eq. (4.3). The results of these calculations (Table 10) illustrate quite well the major tendencies in hydrogel swelling change (exemplified by the crosslinked PAAm) resulting from the effect of pressure and the ionic strength. Thus, Table 10 shows that the absolute available water reserve in SAH (ΔW) increases with β and x_c. However, the higher the "free" swelling of the hydrogel, the lower is the fraction of this reserve in total swelling. These and similar data can be used as a basis for SAH design or choice in any specific situation.

Undoubtedly, the increasing salt tolerance of "agricultural" hydrogels is a very important task. As far as we know, this problem is far from having "inexpensive" solutions which could meet agricultural economy requirements.

Table 10. Characteristics of hydrogel swelling and efficiency as a soil additive[1] [139]

Swelling parameter	β = 0.02			β = 0.10			β = 0.20	
	1500	3000	6000	800	1500	3000	800	1500
$w_{free}(H_2O)$	282	890	3160	676	2290	9120	1820	6310
w_{free}[2]	112	186	355	303	545	998	676	1230
w (pF = 2)	32	40	117	93	105	112	204	676
Δw(4.2 > pF > 2)[3]	29	37	113	90	101	109	200	672
γ[4]	0.40	0.21	0.11	0.45	0.24	0.11	0.37	0.19
B	0.971	0.981	0.990	0.989	0.993	0.996	0.995	0.997

[1] Calculation by the scheme developed in Refs. [22, 102]; crosslinked PAAm with varied ionization degree (β) and $\bar{x}_c$ given in the heading of the Table.
[2] In 0.01 N aqueous NaCl.
[3] In all the cases w (pF = 4.2) ~ 3–3.5 $g \cdot g^{-1}$.
[4] Coefficients from Eq. (4.3): $\gamma = w_{free}(0.01\ N\ NaCl)/w_{free}(H_2O)$; $B = [w_{free} - w\ (pF = 4.2)]/w_{free}$

4.4 Adjusting of Hydrogels with Soil Conditions

The previous analysis of SAH behavior in the soil clearly shows that their application for improving the water-holding capacity is not universal. Hydrogel swelling in a porous, partially salinized medium is affected by numerous factors, most often negative, and therefore a rational application of SAH demands an accurate consideration of these factors. It is evident that certain principles for adjustment of hydrogels to physical and chemical soil parameters, as well as appropriate laboratory tests and calculation algorithm systems should be worked out.

Referring to the ionic effects, measuring of swelling in solutions which closely model real ones can provide reliable estimates. The papers [58, 132] can serve as examples of such an approach. In choosing a type of SAH suitable for some particular soil it is necessary to take into account the acid-base properties of the gel and the soil because otherwise collapse phenomena are likely to result from common counterions and the sorption on solid surfaces.

Various effects of pressure on the SAH swelling have already been discussed above. The universal character of the $\pi(w)$ or $pF(w)$ curves (Figs. 5, 8) would seem to give no chance for the hydrogel "to get rid" of the pressure in order to preserve the same high swelling in the porous medium as in a free state. The ideal situation can be imagined as follows. Each SAH particle occupies a "niche" of its own and reaches its walls upon swelling. In this case, the SAH effect is likely to be additive, and then Eq. (4.2) should be true. Johnson [58] described this unique kind of behavior; the PAAm hydrogel with swelling of $125\ g \cdot g^{-1}$ in hard tap water produces a linear increase in the water content of coarse sand with an introduced dose, its swelling in the soil medium being equal to $121\ g \cdot g^{-1}$.

However, in most cases the $\Delta W(D)$ dependencies are distinctly nonlinear (Fig. 9), which gives impulse to further speculations. Clearly, dependencies of this type can result only from mutual suppression of the hydrogel particles because of their nonuniform distribution over the pores as well as from the presence of a distribution with respect to pore size which does not coincide with the size distribution of the SAH swollen particles. A considerable loss in swelling followed from the $W(D)$ dependencies, as shown in Fig. 9, need a serious analysis which most probably would lead to the necessity of correlating the hydrogel particle sizes with those of the soil pores as well as choice of the technique of the SAH mixing with the soil. Attempts to create the appropriate mathematical model have failed, for they do not give adequate results.

The dynamic aspect of the SAH behavior is also connected with the particle sizes. Indeed, the hydrogel swelling and the loss of water in the soil proceed in a pulse regime, the frequency of pulses being random (rainfalls) or regular (irrigation). The relaxation times for swelling, deswelling, and water evaporation, which are all diffusion processes, should increase as the square of the particle diameter (see Sect. 3.3). Here, some contradictory requirements to the particle sizes emerge, for the swelling rate and the correlation with the pore sizes demand their decrease, while the need to prolong water retention requires the opposite, i.e., to enlarge the particle sizes. Problems of this kind should be solved with respect to specific or typical situations and the algorithms for their solution is more or less clear.

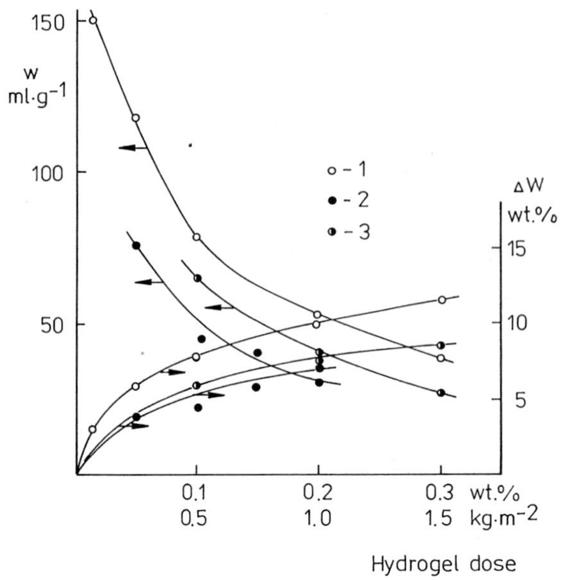

Fig. 9. Dose dependencies of the hydrogel swelling and the increase in water-holding capacity of sands amended by hydrogels; *1* — the Lower Volga bank [133], *2* — the Kara Kum desert [132], *3* — the Inshas area, Egypt [10]

The entire additional water content in the soil created by SAH depends on its lifetime or the limiting number of swelling-deswelling cycles, which is reflected in Eq. (4.3). These problems are scarcely discussed in the literature. However, it is possible to assume that both physical and biological factors would limit the SAH lifetime in the soil. Among the biological ones, attention should be paid to the attack of soil bacteria. The linear PAAm as a soil conditioner is known to exhibit a sufficiently high bacterial stability [138], while in the case of polymers containing natural components adequate protection seems to be necessary.

As to the physical factors, the swelling pressure in course of multiply repeating cycles, and the temperature effects — freezing the water inside swollen particles — should result in degradation of the hydrogel network structure. No data allowing any prediction are available in the literature so far, though some companies indicate that hundreds of swelling cycles are possible without any substantial changes in the SAH ability to accumulate water. Direct long-term experiments seem to be necessary.

5 Conclusion

The hydrophilic network polymers have acquired a completely new field of application, the word "field" being understood here literally. SAH as a new soil water-retaining agent to be used in water-deficit conditions is attracting widespread attention, which is manifested in a great number of publications and the expanding production of corresponding materials.

Physical foundations of SAH application seem more or less clear and have been confirmed experimentally. As for the biological efficiency in solving plant growth problems and, especially, the economic feasibility of their application, the available information is far from being sufficient, though it is obvious that the use of SAH can open new ways in controlling water, fertilizer, and pesticide consumption as well as to enhance crop production in arid zones. The principle "More water with a lower dosage" stimulates further development of SAH with optimal properties for specific situations.

The thermodynamic point of view developed in this review and in our original works with regard to the behavior of SAH in laboratory experiments and in soil models can pave, in our opinion, the most rational way for achieving the optimal results. Based on the existing theory of network polymers, this concept is undoubtedly open to further improvement that would expand its prognostic potentialities.

6 References

1. Wichterle O, Lim D (1960) Nature 185: 117
2. Andrade JD (ed) (1976) Hydrogels for medical and related applications. ACS symp ser 31. ACS, Washington
3. Peppas NA (ed) (1986) Hydrogels in medicine and pharmacy. CRC Press, Boca Raton
4. Flory PJ (1953) Principles of polymer chemistry. Cornell Univ Press, Ithaca, New York
5. Masuda F (1983) Chem Econ Eng Rev 15(11): 19
6. Levy M, Vofsi D (1982) ACS Polym Prepr 23: 197
7. Jensen MH, King PA, Eikhof R (1971) In: 10th Natl Agric Plast Conf Proc Chicago, p 69
8. Eikhof RH, King PA, Moline W (1973) In: 11th Natl Agric Plast Conf Proc San Antonio, p 117
9. King PA, Eikhof RH, Jensen MH (1973) ibid., p 106
10. El-Hady OA, Azzam R (1983) Egypt J Soil Sci 23: 243
11. Azzam R et al. (1983) In: Proc Intl Symp on Isot and Rad Tech in Soil Phys and Irrig Stud Aix-en-Provance, p 321
12. Azzam R (1985) Commun Soil Sci Plant Anal 16: 1123
13. Azzam R, Danial L, Ayoub R (1986) In: Proc 6 Tihany Symp Rad Chem Siofok, 2: 649
14. Hemphill DD (1982) Hort Sci 17: 256
15. Ward SM, O'Driscoll F (1985) Agr Mechaniz in Asia, Afr and Lat Amer 16: 45
16. Gaal O, Medgyesi GA, Vereczkey L (1980) Electrophoresis in the separation of biological macromolecules. Akadémiai Kiado, Budapest
17. Tanaka T (1979) Polymer 20: 1404
18. Baselga J et al. (1987) Macromolecules 20: 3060
19. Ilavsky M (1982) Macromolecules 15: 782
20. Geissler E et al. (1988) Macromolecules 21: 2594
21. Oppermann W, Rose S, Rehage G (1985) Br Polym J 17: 175
22. Dubrovskii SA et al. (1990) Polym Bull 24: 107
23. Truong ND et al. (1986) Polymer 27: 459, 467
24. Baselga J et al. (1989) Polym J 21: 467
25. Watkin JE, Miller RA (1970) Anal Biochem 34: 424
26. Gupta MK, Bansil R (1981) J Polym Sci: Polym Phys Ed 19: 353
27. Bansil R, Gupta MK (1980) Ferroelectrics 30: 63
28. Hsu TP, Cohen C (1983) Polymer 24: 1273
29. Hsu TP, Ma DS, Cohen C (1984) Polymer 25: 1419
30. Janas VF, Rodrigues F, Cohen C (1980) Macromolecules 13: 977
31. Weiss N, Silberberg F (1977) Br Polym J 9: 144
32. Nieto JL et al. (1987) Eur Polym J 23: 551
33. Baselga J et al. (1988) Eur Polym J 24: 161
34. Baselga J et al. (1989) Eur Polym J 25: 471, 477
35. Rogozhin SV, Vainerman ES, Lozinsky VI (1982) Dokl Akad Nauk SSSR 263: 115
36. Lozinsky VI et al. (1986) Acta Polym 37: 142

37. Thomson RAM (1983) In: Finch CA (ed) Chemistry and technology of water-soluble polymers. Plenum Press. New York, p 31
38. Kulicke WM, Kniewske R, Klein J (1982) Progr Polym Sci 8: 373
39. Huang MY, Heng SY (1983) Makromol Chem, Rapid Commun 4: 17
40. Rosiak J, Burczak K, Czolozińska T (1983) Rad Phys Chem 22: 917
41. Buyanov AL, Revelskaya LG, Petropavlovskii GA (1989) Zh Prikl Khim 62: 1854
42. Buyanov AL et al. (1989) Vysokomol Soedin, Ser B 31: 883
43. Fanta GF et al. (1982) Starch 34: 95
44. Fanta GF et al. (1978) Starch 30: 237
45. Castel D, Ricard A, Andebert R (1988) J Macromol Sci, Chem 25: 235
46. Castel D, Ricard A, Andebert R (1990) J Appl Polym Sci 39: 11
47. Fanta GF, Burr RC, Doan WM (1979) J Appl Polym Sci 24: 2015
48. Kazanskii KS et al. (1988) Vestn S-H Nauki 4: 125
49. Vitta SB, Stahel EP, Stannett VT (1985) J Macromol Sci, Chem 22: 579
50. Fanta GF, Burr RC, Doane WM (1982) J Appl Polym Sci 27: 4239
51. Taylor NW et al. (1978) J Appl Polym Sci 22: 1343
52. Finch CA (1983) in: Ref. [37] p 81
53. Horkay F, Nagy M (1981) Acta Chim Hung 108: 111
54. Horkay F, Nagy M (1982) Acta Chim Hung 109: 415
55. Peppas NA, Benner RE (1980) Biomaterials 1: 158
56. Reinhardt CT, Peppas NA (1984) J Membr Sci 18: 227
57. Mateescu MA et al. (1984) Polym Bull 11: 421
58. Johnson MS, (1984) J Sci Food Agric 35: 1063, 1196
59. Westman L, Lindstrom T (1981) J Appl Polym Sci 26: 2519
60. Taylor NW (1979) J Appl Polym Sci 24: 2031
61. Kobayashi S et al. (1989) Polym J 21: 971
62. Chujo Y et al. (1989) Macromolecules 22: 1074
63. Hunt JA et al. (1989) AIChE J 35: 250
64. Allain C, Salomé L (1987) Polym Commun 28: 109
65. Prud'homme RK et al. (1983) Soc Pet Eng J 23: 804
66. Jordan DS et al. (1982) Soc Pet Eng J 22: 463
67. Allain C, Salomé L (1988) In: Kramer O (ed) Biological and synthetic polymer networks. Elsevier, New York, p 291
68. Allain C, Salomé L (1987) Macromolecules 20: 2957
69. Rudman AR et al. (1983) Vysokomol Soedin, Ser A 25: 2405
70. Kopylova YeM et al. (1987) Vysokomol Soedin, Ser A 29: 517
71. Watase M et al. (1983) Polym Commun 24: 52, 270, 345
72. Fanta GF, Weaver MO, Doane WM (1974) Chem Tech 4: 675
73. Chapiro A (1960) Radiation chemistry of polymeric systems Wiley-Interscience, New York
74. Marchal J (1965) C r Acad Sci 261: 5104
75. King PA, Ward JA (1970) J Polym Sci A-1 8: 253
76. Ward JA (1971) J Polym Sci 9: 3555
77. Minkova L et al. (1989) J Polym Sci: Polym Phys 27: 621
78. Kazanskii KS, Arkhipovich GN (to be published)
79. King PA (1966) US Pat 3.264.202
80. Burillo G, Ogawa T (1980) Makromol Chem Rapid Commun 1: 545
81. Burillo G, Ogawa T (1981) Rad Phys Chem 18: 1143
82. Burillo G, Ogawa T (1986) J Appl Polym Sci 32: 3783
83. Buchanan KJ, Hird B, Letcher TM (1986) Polym Bull 15: 325
84. Chapiro A, Legris C (1986) Rad Phys Chem 28: 143
85. Ikada Y et al. (1977) Rad Phys Chem 9: 633
86. Rosiak J, Olejniczak J, Charlesby A (1988) Rad Phys Chem 32: 691
87. Yen SW, Osterholz FD (1975) US Pat 3.900.378
88. Assarsson PG, King PA (1976) US Pat 3.898.443, 3.957.605, 3.993.551, 3.993.552
89. Yoshikawa M et al. (1989) New Polym Mat 1: 223
90. Gnanou Y, Hild G, Rempp P (1984) Macromolecules 17: 945

91. Graham NB, Zulfigar M (1989) Polymer 30: 2130
92. Rogers JA, Tam T (1977) Can J Pharm Sci 12: 65
93. Graham NB, McNeil ME (1988) Makromol Chem, Makromol Symp 19: 255
94. Muller G, Laine JP, Fenyo JC (1979) J Polym Sci: Polym Chem Ed 17: 659
95. Kulicke WM, Hörl HH (1985) Colloid Polym Sci 263: 530
96. Tanaka T (1987) In: Nicolini C (ed) Structure and dynamics of biopolymers. Nijhoff Publ, Dordrecht, p 237 (NATO ASI Series E, No 133)
97. Tanaka T et al. (1980) Phys Rev Lett 45: 1636
98. Ilavský M (1981) Polymer 22: 1687
99. Vasilevskaia VV, Khokhlov AR (1986) Vysokomol Soedin, Ser A 28: 316
100. Harrison DJP, Yates WR, Johnson JE (1985) J Macromol Sci Rev Macromol Chem Phys 25: 481
101. Rička J, Tanaka T (1984) Macromolecules 17: 2916
102. Dubrovskii SA et al. (1989) Vysokomol Soedin, Ser A 31: 321
103. Dubrovskii SA et al. (1990) Vysokomol Soedin, Ser A 32: 165
104. Horkay F, Zrinyi M (1982) Macromolecules 15: 1306
105. Candau S, Bastide J, Delsanti M (1982) Adv Polym Sci 44: 27
106. Nossal R (1985) Macromolecules 18: 49
107. Rička J, Tanaka T (1985) Macromolecules 18: 83
108. Kulicke WM, Nottelmann H (1987) Polym Mater Sci Eng 57: 265
109. Nishi S, Kotaka T (1986) Macromolecules 19: 978
110. Bailey FE, Koleske JV (1976) Polyethylene oxide. Academic Press, New York
111. Khokhlov AR (1980) Polymer 21: 376
112. Nagy M, Horkay F (1979) Magy Kem Foly 85: 513
113. Hasa J, Ilavsky M, Dušek K (1975) J Polym Sci Polym Phys Ed 13: 253
114. Koňák Č, Bansil R (1989) Polymer 30: 677
115. Horkay F, Hecht AM, Geissler E (1989) Macromolecules 22: 2007
116. Dušek K, Prins W (1969) Adv Polym Sci 6: 1
117. Khokhlov AR (1980) Vysokomol Soedin, Ser B 22: 736
118. Mark JE (1982) Adv Polym Sci 44: 1
119. De Gennes P-G (1979) Scaling concepts in polymer physics. Cornell Univ Press, Ithaca, New York
120. Ilavský M, Hrouz J (1982) Polym Bull 8: 387
121. Ilavský M, Hrouz J (1983) Polum Bull 9: 159
122. Watase M (1985) Makromol Chem 186: 1081
123. Starodubtsev SG (1982) Vysokomol Soedin, Ser B 24: 67
124. Brannon-Peppas L, Peppas NA (1988) Polym Bull 20: 285
125. Tanaka T, Fillmore DJ (1979) J Chem Phys 70: 1214
126. Sato Matsuo E, Tanaka T (1988) J Chem Phys 89: 1695
127. Ilmain F, Candau SJ (1989) Makromol Chem Makromol Symp 30: 119
128. Dubrovskii SA (1988) Dokl Akad Nauk SSSR 303: 1163
129. Tanaka T et al. (1987) Nature 325: 796
130. Sekimoto K, Kawasaki K (1989) Physica A 154: 384
131. El-Hady OA et al. (1981) Acta Hort 119: 247, 257
132. Nuriev BN et al. (1986) in: Monakov VS (ed) Technical progress in deserts (in Russian). Ilym, Ashkhabad, p 59
133. Sus NS et al. (1990) Pochvovedenije 7: 149
134. Garrison S (1981) The thermodynamics of soil solutions. Clarendon Press, Oxford
135. Voronin AD (1984) Structural functional hydrophysics of soil (in Russian). Moscow Univ Press, Moscow
136. Yoshitake T (1981) Polym Digest 33: 10
137. Stevenson DS (1987) Can J Soil Sci 67: 395
138. Grula MM, Huang M (1982) Dev Ind Microbiol 22: 451
139. Lagutina MA et al. (to be published)

Editor: K. Dušek
Received July 1, 1991

Polymer-Coated Adsorbents for the Separation of Biopolymers and Particles

A. E. Ivanov, V. V. Saburov and V. P. Zubov
Shemyakin Institute of Bioorganic Chemistry, Russian Academy of Sciences, ul. Miklukho-Maklaya 16/10, Moscow, Russia

Polymer-coated mineral supports are materials attracting increasing interest in the field of bioseparation. Adsorbents of this type often combine the rigidity of silica, the biocompatibility of organic gels and/or the chemical stability of bulky polymers. The review collects the methods of preparation, structures of the stationary phases as well as probable mechanisms of their interaction with biomacromolecules. It was our aim to analyze the interrelation between the structural and chromatographic characteristics of composite adsorbents. Up to now specific properties of the surface-covering polymers (such as an excluded volume effect, segment dynamics, etc.) have hardly come under the scruting of chromatography researchers. Recently, a number of applications have confirmed that the value of the polymer structure should be taken into accounting. In this paper both hydrophilic diffuse polymer layers and condensed thin hydrophobic films are considered as interfaces of the adsorbing media. The most promising types of bioseparation procedures are given for each type of adsorbent.

Advances in Polymer Science, Vol. 104

1 Introduction

The modern requirements for properties of "an ideal chromatographic sorbent" especially for the separation of biomolecules and particles are rather extensive out and comprise a few controversial features which can hardly ever be achieved in a single material. Those include:

- insolubility;
- permeability to macromolecules;
- high rigidity, well defined porosity independent of solvent;
- large specific surface area;
- low non-specific adsorptivity;
- physico-chemical and biological stability;
- facile derivatization.

Modern trends in development of organic polymer sorbents with semi-rigid porous structure or of rigid inorganic matrices with various bonded phases (based on organosilicon low molecular weight modifiers) to some extent provide a fairly good compromise of the sorbent properties. However, in our opinion, the best possibility of combining the rigidity and excellent pore-size distribution of silicas with biocompatibility and facile derivation of soft organic gels may be expected by creation of a material composed of an inorganic skeleton modified by adsorbed or grafted macromolecules ("composite sorbent"). The pioneering investigations of Regnier et al., Alpert, Jozefonvicz et al. and some others, including the authors of this paper, have shown the considerable potential of these materials for high-performance liquid chromatography of proteins and nucleic acids. This field was reviewed by us in 1989 [1]. In the present review we are trying not only to update the intensively developing area of composite sorbents but also to discuss some basic problems of structure and behavior of polymeric bonded phases. These include:

- the interrelationship between the structure and chromatographic properties of the sorbents,
- the role of excluded volume effects and chain dynamics in the behavior of polymeric bonded phases,
- the influence of macromolecular spacer arms on bioaffinity separations.

We believe that better understanding of the behavior of macromolecules at the solid surfaces will facilitate further progress in chemical design of the composite sorbents as well as other bioseparation media such as membranes, fibers etc. and their application in various fields.

2 Theoretical Background

2.1 Passivity Mechanism

Numerous applications of polymer-coated silicas to chromatography of biopolymers allow one to conclude that adsorbed or grafted hydrophilic nonionizing macromolecules in many cases supply the silica matrix with the remarkable repellency towards proteins and nucleic acids displayed by moderate salt concentrations in contacting buffers. The inherent adsorptivity of silicas is suppressed by the attached polymers, moreover, the properties of the „inert" adsorbent may be modified by the introduction of charged or hydrophobic groups into the polymer.

The qualitative thermodynamic explanation of the shielding effect produced by the bound neutral water-soluble polymers was summarized by Andrade et al. [2] who studied the interaction of blood with polyethylene oxide (PEO) attached to the surfaces of solids. According to their concept, one possible component of the passivity may be the low interfacial free energy (γ_{sl}) of water-soluble polymers and their gels. As estimated by Matsunaga and Ikada [3], it is 3.7 and 3.1 mJ/m^2 for cellulose and polyvinylalcohol whereas 52.6 and 41.9 mJ/m^2 for polyethylene and Nylon 11, respectively. Ikada et al. [4] also found that adsorption of serum albumin increases dramatically with the increase of interfacial free energy of the polymer contacting the protein solution.

Absolom et al. [5] considered protein adsorption from solutions with different surface tensions (γ_{lv}). The authors studied the adsorption of several serum proteins (immunoglobulins IgG and IgM, human serum albumin, a_2-macroglobulin) from phosphate buffered saline (pH 7.3, $\gamma_{lv} = 72.9$ mJ/m^2) and dimethylsulphoxide-containing water solutions with lower surface tensions ($\gamma_{lv} = 63.2$; 67.2; 69.1 mJ/m^2) on the substrate materials exhibiting a wide range of surface tensions (γ_{sv}) from 17 to 67 mJ/m^2. The plateau level of adsorption ($\mu g/cm^2$) for each protein and each of the substrates has been plotted as a function of substrate surface tensions. With increasing surface tension of the solid substrate (γ_{sv}) or, what is the same, with decreasing interfacial free energy (γ_{sl}) the plateau level of adsorption decreases in experiments carried out in a phosphate buffer.

In the absence of specific interactions of the receptor – ligand type the change in the Helmholtz free energy (ΔF_{ads}) due to the process of adsorption is $\Delta F_{ads} = \gamma_{ps} - \gamma_{pl} - \gamma_{sl}$, where γ_{ps}, γ_{pl} and γ_{sl} are the protein-solid, protein-liquid and solid-liquid interfacial tensions, respectively [5]. It is apparent from this equation that the free energy of adsorption of a protein onto a surface should depend not only of the surface tension of the adhering protein molecules and the substrate material but also on the surface tension of the suspending liquid. Two different situations are possible.

1. For $\gamma_{lv} > \gamma_{pv}$, where γ_{lv} and γ_{pv} are the surface tensions of liquid and protein, respectively, ΔF_{ads} increases with increasing γ_{sv}, predicting decreasing polymer adsorption. An example of this is phosphate buffer saline where $\gamma_{lv} = 72.9$ mJ/m^2 and γ_{pv} is usually between 65 and 70 mJ/m^2 for most proteins [5]. Therefore, supports for gel-permeation and affinity chromatography should be as hydrophilic as possible in order to minimize undesirable adsorption effects.

2. For $\gamma_{lv} < \gamma_{pv}$ the opposite behavior pattern is predicted, namely ΔF_{ads} decreases with increasing γ_{sv}, thus stronger adsorption occurs. Indeed, adsorption of IgM from the dimethylsulphoxide solutions in water ($\gamma_{lv} = 63.2$; 67.2; 69.1) increases with increasing substrate surface tension (γ_{sv}).

When speculating about the hypothetical structure of interfaces with minimal free energy, the diffuse interfaces formed by the properly grafted water-soluble polymers were predicted to be ideal [6]. The diffuse interface may contain a large amount of water and a small number of polymer chains covalently bound to the substrate materials with good mechanical properties. It is likely that this two-phase system has a γ_{sl} value close to zero.

Another component of the shielding effect provided by adsorbed or grafted polymers are the conformational changes in the bound macromolecular coil which is in contact the protein globule. Since the coils obviously have extended loops and tails as previously suggested [7], the repulsive forces are generated by the loss of possible chain conformations, as the volume available to the adsorbed chains is reduced between the approaching surfaces [6]. Besides, the rotational mobility of pendant groups (if present) may be restricted so that the overall entropy of the system is decreased. The enthalpic unfavorability of the inner chain contacts in a good solvent (water) also contributes to the free energy of the system [8]. As a response, the macromolecules bound to the surface will reorient to their local microenvironment so that the approaching protein globules will be repulsed.

We think, therefore, that the conformation, chain and segment mobilities in the attached macromolecules can play a significant role in the shielding behavior of the polymeric stationary phase as well as in the processes of its formation of complexes with solutes. Obviously, the chromatographic studies relevant to composite supports suffer from a lack of information on the structure of the attached polymer. Nevertheless, we will attempt to point out some relevant data from independent studies on polymer adsorption and/or graft polymerization.

2.2 Theories of Polymer Adsorption

Adsorption of macromolecules has been widely investigated both theoretically [9–12] and experimentally [13–17]. In this paper our purpose was to analyze the probable structures of polymeric stationary phases, so we shall not go into complicated mathematical models but instead consider the main features of the phenomenon. The current state of the art was comprehensively summarized by Fleer and Lyklema [18]. According to them, the reversible adsorption of macromolecules and the structure of adsorbed layers is governed by a subtle balance between energetic and entropic factors. For neutral polymers, the simplest situation, already four contributor factors must be distinguished:

(a) the segmental adsorption energy;
(b) the chain conformation entropy;
(c) the entropy of mixing of segments and solvent;
(d) interactions between segments (and solvent).

Part (a) is the driving force for the adsorption. If only (a) were present, adsorbed chains would lie flat on the surface. Parts (b) and (c) are the opposing forces: (b) accounts for the entropy loss of a bond on the surface as compared to the solution, (c) represents the separation into a concentrated surface "phase" and a dilute solution. Part (d) arises from polymer-polymer, solvent-solvent and polymer-solvent interactions, which usually favour accumulation of segments. At equilibrium at a finite solution concentration, (a)–(d) lead to extended layers with protruding (short) loops and (long) tails.

For flexible polyelectrolytes, additional electric contributions occur. If the surface is charged, (a) may be higher or lower, depending on the charge signs. A polyelectrolyte adsorbs on a surface with a like charge only if the "chemical" adsorption energy exceeds the electrostatic repulsion between segments and surface. Parts (b) and (c) remain, in principle, the same. Contribution (d) is affected strongly since electrostatics oppose accumulation of charges.

Moreover, this repulsion now has a longer range, depending on the ionic strength. In solutions of low salt concentration the effect of (d) is most pronounced. The strong tendency of segments to avoid each other is only overcome in the surface layer (when (a) is high enough). Consequently, adsorbed layers in salt-free solutions are flat with only a small number of tails sticking out.

This model, however, has been reappraised by Barford et al. [19, 20]. They supposed a non-equilibrium type of interaction between the surface and a polyelectrolyte macromolecule (the binding energy per segment might be many times kT). According to their model [19], an incremental amount of polymer is initially adsorbed onto a surface. Once it has been adsorbed, its concentration profile is fixed. Another incremental amount of polymer is adsorbed onto a surface, but its adsorption is hindered by the repulsive excluded volume potential of the adsorbate. Since the adsorbed macromolecules are anchored and the shape of loops is thought to be unchanged, the polymer system can not minimize its free energy so far as it occurs in the case of equilibrium adsorption. In comparison with the latter, the less polymer can be adsorbed while the polymer profiles are more extended, as a function of adsorbate concentration, in non-equilibrium adsorption. One can expect, therefore, that the diffuse structure with longer tails is formed as a result of irreversible adsorption of macromocules, both ionizable and neutral [19].

These predictions are in better agreement with the results of Luckham and Klein [21] who observed an extended force profile between two surfaces coated with polyelectrolyte and supposed the non-equilibrium character of polymer adsorption.

In the following paper, the possibility of equilibration of the primarily adsorbed portions of polymer was analyzed [20]. The surface coupling constant (k_0) was introduced to characterize the polymer-surface interaction. The constant k_0 includes an electrostatic interaction term, thus being $k_0 > 1$ for polyelectrolytes and $k_0 \ll 1$ for neutral polymers. It was found that, theoretically, the adsorption characteristics do not depend on the equilibration processes for $k_0 > 1$. In contrast, for neutral polymers ($k_0 \ll 1$), the difference between the equilibrium and non-equilibrium modes could be considerable. As more polymer is adsorbed, excluded-volume effects will swell out the loops of the adsorbate, so that the mutual reorientation of the polymer chains occurs.

Another important feature of polymer adsorption is the influence exerted on it by the surface roughness. Ball et al. [22] proposed that if the surface potential is not attractive enough to bind the polymer when flat, then corrugation can aid binding as follows. For a sinusoidal corrugation, one might anticipate that some parts will become locally more favorable to bind and other less: the polymer will than preferentially occupy the favorable regions and net binding can result. In other words, after adsorption (contribution (b), according to [18]), the loss of the chain entropy can be smaller than that for a flat surface. Also, it can be overcome by the greater gain in the segmental adsorption energy, "(a) contribution" [18], enhanced by the proper orientation of the macromolecule near the surface.

The results obtained demonstrate competition between the entropy favouring binding at bumps and the potential most likely to favour binding at dips of the surface. For a range of pairwise-additive, power-law interactions, it was found that the effect of the potential dominates, but in the (non-additive) limit of a surface of much higher dielectric constant than in solution the entropy effects win. Thus, the preferential binding of the polymer to the protuberances of a metallic surface was predicted [22]. Besides, this theory indirectly assumes the occupation of bumps by the weakly attracted neutral macromolecules capable of covalent interaction with surface functions.

Adsorption of polymers on the rough surfaces was considered by Douglas in terms of fractals [23]. According to his study, adsorption occurs more readily on fractal surfaces since it requires a smaller "entropic price". Concretely, the free energy of adsorption per monomer unit of the polymer chain is proportional to $|(T - \Theta_A)/T|^{-1/\varphi_s}$, where $\varphi_s = (d_{fc} + d_{fs} - d)/d_{fc}$; d_{fc}, d_{fs} and d — fractal dimensions of the chain, surface and space (d = 3), Θ_A is a critical temperature of adsorption and T — sample temperature. As a result, adsorption of polymer tends to increase with d_{fs}. Therefore, a rough surface may adsorb a polymer when a corresponding smooth surface of the same material will not adsorb. Roughness can thus alter the effective polymer-surface interaction in a fundamental way. Roughening (or increase of the fractal dimension, d_{fs}) leads to a "tighter" adsorption onto a surface (i.e. decreases the thickness of the polymer layer relatively to the ideal planar surface). Besides, adsorption will apparently increase with d_{fc}. In other words, flexible chain macromolecules will adsorb more readily than those bearing a rigid backbone.

The theory of Douglas [23] leads to a better understanding of the qualitative non-equilibrium effects of polymer adsorption. For a strongly attractive plane surface some portion of the polymer will deposit rapidly onto a surface thereby "modifying" the surface on which the remaining part of the chain adsorbs. Such a modified "stringly" surface is akin to a surface pierced with many holes and the effective dimension of the surface can be expected to be lower than that of the starting planar surface. A lower value of d_{fs} leads to a more diffuse interface for the initial non-equilibrium adsorption, which should gradually relax to a more strongly adsorbed configuration through thermal fluctuations. These predictions are in a good agreement with the results of Barford et al. listed above [19, 20].

2.3 Selected Structural Data on Physical Adsorption of Polymers

In addition to theoretical considerations, some experimental studies were carried out to clarify the structures of adsorbed layers of hydrophilic macromolecules.

Hommel et al. [24] have found that the conformation of polyethylene oxide chains grafted to silica depends strongly on the grafting ratio characterized by the surface area occupied by a single macromolecule. In this study, no exact evidence of the chemical binding was obtained, moreover, adsorption and end-group coupling both seem to be possible because the „grafting" of PEO was carried out at elevated temperature (113 °C or 160 °C) by the evaporation of an aqueous solution.

Grafted PEO layers in contact with water have been studied using a spin label bound to the free end of the grafted polymer of molecular weight 1000 (PEG-1000; N = 23). It was shown that at low surface concentration of PEO (S = 0.17 molecule/nm^2 or 0.28 mg/m^2) the labeled polymer exists in two motional domains. A portion moving slowly is relevant to the labels that are strongly interacting with polar silanols on the non-bound region of the surface and a fast domain derives from the labeled tails in the solvent. When the grafting ratio was about twice as high (S = 0.37 molecule/nm^2 or 0.61 mg/m^2), only the fast domain remained detectable. Apparently, the chains spread into solution: they interact, repel each other and adopt a more extended configuration.

In contrast to PEG-1000, the ESR-spectra obtained for labelled grafted PEO-oligomers (N = 1–9) were reported to differ strongly from that for N = 23 and exhibited only slowly moving labels. As the line width of the fast domain spectrum depends on the mobility of the label, it can be concluded that the tails for N = 23 must consist of at least 4–9 repeated units. It appears that some tails are rather long.

A very similar effect of the surface concentration on the conformation of adsorbed macromolecules was observed by Cohen Stuart et al. [25] who studied the diffusion of the polystyrene latex particles in aqueous solutions of PEO by photon-correlation spectroscopy. The thickness of the hydrodynamic layer δ (nm) calculated from the loss of the particle diffusivity was low at low coverage but showed a steep increase as the adsorbed amount exceeded a certain threshold. Concretely, δ increased from 40 to 170 nm when the surface concentration of PEO rose from 1.0 to 1.5 mg/m^2. This character of the dependence is consistent with the calculations made by the authors [25] according to the theory developed by Scheutjens and Fleer [10, 12] which predicts a similar variation of the hydrodynamic layer thickness of adsorbed polymer with coverage. The dominant contribution to this thickness comes from long tails which extend far into the solution.

An important feature of physically adsorbed macromolecules is their exchangeability with other macromolecules that exist in a contacting solution. It has been shown by several authors [16, 26] that in a polydisperse sample, preferential adsorption of large molecules over smaller ones occurs. Especially, with increasing molecular weight the polymers adsorb progressively more strongly and can replace molecules of smaller M. Therefore, the molecular weight distribution in the adsorption layer changes with the amount of adsorption and may differ considerably from that in the bulk phase.

Furusawa and Yamamoto [16] studied the adsorption process of polystyrene samples (M ranging from 16700 to 2×10^6) with narrow molecular weight distribution (M_w/M_n = 1.01–1.07) at the Θ-conditions (cyclohexane, 35 °C). Controlled pore glass with pore diameter of 1000 Å was used as an adsorbent.

For the mixture of 186000 and 775000 polystyrenes (1:1) the overall adsorption reached an equilibrium within 1 h, but the relative change of each component in the adsorption layer continued for a long time (24 h). The final composition was enriched by the larger macromolecules (3:1, 24 h), as compared with the beginning of the process (1:1, 0.1 h). The exchangeability of the adsorbed polymer layer was found to depend strongly, not only on the molecular weight of the adsorbed polymer, but also on the solution concentration at which the corresponding initial adsorption was carried out. It appears that polymer molecules adsorbed at high concentrations are exchanged more easily than molecules adsorbed at low concentrations. If the exchangeability is related to the number of adsorbed segments, this suggests that the polymer layer formed at high concentrations has fewer adsorbed segments than the layer formed at low concentrations. This can be explained by the polymer molecules in a dilute solution arriving at the sparsely populated surface and adopting a relatively flat conformation. As calculated by Cosgrove [27], the isolated chain is almost totally unfolded on the surface if its binding energy exceeds 2 kT per monomer unit. However, the polymer molecules in a more concentrated solution, especially when some later arrivals come to the surface, would occupy a smaller number of sites and therefore must adopt a considerably extended loop conformation which would be desorbed more easily than the flat ones.

This exchangeability of adsorbed layers should be considered for better understanding of the irreversible adsorption of polymers. Apparently, penetration by the macromolecules adsorbed later through the layer of the initially adsorbed ones will include a slow exchange between the positions of segments and take a longer time.

The consideration made above allows us to predict good chromatographic properties of the bonded phases composed of the adsorbed macromolecules. On the one hand, steric repulsion of the macromolecular solute by the loops and tails of the modifying polymer ensures the suppressed nonspecific adsorptivity of a carrier. On the other hand, the extended structure of the bonded phase may improve the adaptivity of the grafted functions and facilitate thereby the complex formation between the adsorbent and solute. The examples listed below illustrate the applicability of the composite sorbents to the different modes of liquid chromatography of biopolymers.

3 Chromatographic Packings Prepared by Physical Adsorption of Polymers on Silica or Porous Glass

3.1 Adsorption of Neutral Polymers

It has been outlined by several authors that the single macromolecule may be irreversibly bound because of the large number of weakly interacting segments. The first papers on the construction of polymer-coated silica adsorbents involved the physical adsorption of water-soluble polymers. Polyethylene oxides [28, 29] and poly-*N*-vinylpyrrolidone [30] are examples of the stationary phases of this type.

In 1971, Hiatt et al. found that polyethylene oxide (PEO) of molecular weight about 100000 prevented the adsorption of rabies virus to porous glass with an average pore diameter of 1250 Å. The support was modified by passage of one void volume of 0.4% solution of the polymer in water, followed by 5 or more volumes of distilled water or buffered salt solution. The virus was effectively purified from the admixtures of brain tissue fluid by means of size-exclusion chromatography on the modified glass column [28].

Porous glass (1250 Å, 80–120 or 120–200 mesh) coated with polyethylene glycol (M = 20000) was used for preparative isolation of the avian myeloblastosis virus from chicken plasma and the hamster melanoma virus from tissue culture fluid by means of size exclusion chromatography [29]. Electron micrography of the purified virus particles revealed that its fine structure had not been damaged. 1100-fold purification from the plasma proteins had been achieved. When 363 Å pore-size glass beads were treated with polyethylene glycol they were able to be used for size-exclusion chromatography of proteins, and this was proved with a series of proteins of varying molecular weight. The RNA-directed DNA-polymerase was separated from other viral components using this chromatographic column. However, retardation of basic proteins (pI > 7.5) was observed assumably due to interaction with the glass surface.

Chromatographic properties of polyvinylpyrrolidone-coated silica (PVP-silica) were studied in order to check whether it could be used in size-exclusion chromatography of water soluble polymers including proteins and dextrans [30]. Isotherms found for PVP (M = 10000) adsorption on LiChrosphere packings with pore sizes 500, 1000 and 4000 Å indicated that the same adsorbance per surface area (1 mg/m^2) had been reached for the three silicas at equilibrium concentrations of polymer exceeding 0.05 mg/ml. In contrast, the narrow pore LiChrospher (100 Å) adsorbed only 0.8 mg PVP/m^2 whereas the saturation occurred at concentrations higher than 0.5 mg/ml. The two universal calibration curves (log M[η] vs V_{el}) for dextrans and PEO were identical and proved the absence of non-steric effects in the elutions from the wide pore packings. Alternatively, adsorption occurred with PEO chromatographed on the 100 Å pore silica. This phenomenon could be explained by incomplete surface coverage. The PVP used for coating had, in solution, a hydrodynamic volume too large to penetrate completely into the pore structure during the coating procedure.

The coating stability was checked by measuring the amount of coated polymer via elemental analysis and thermogravimetry after washing several times with pure water. No desorption could be detected. Unfortunately, the coating was not very stable in a chloroform – methanol (75 – 25% v/v) mixture, the most appropriate medium for packing the efficient columns. Thus, a column packed with bare silica was then modified by eluting with PVP solution. Separation of some protein mixtures (aldolase, ribonuclease, carbonic anhydrase and bovine serum) by means of the 500 Å pore PVP-silica packing has been demonstrated.

Since the preparation of the PEO and PVP silicas was carried out under the circumstances corresponding to the plateau part of isotherms, it obviously led to tailed structures of the stationary phases. Their inherent repellency ensured the size-exclusion mechanism for chromatography of viruses and large proteins.

Besides, in the absence of long-range attractive interactions, the binding of a neutral polymer at the outward extremities of the surface is more likely [22]. Therefore, these irregularities can be masked by the polymer so that the probability of their interference into the biopolymer structure is diminished. This effect is especially important for chromatography of viruses and cell organelles which are large and more fragile than proteins or oligonucleotides.

The drawback of the described adsorbents is the leakage of the bonded phase that may occur after the change of eluent or temperature of operation when the equilibrium of the polymer adsorption is disturbed. In order to prepare a more stable support Dulout et al. [31] introduced the treatment of porous silica with PEO, poly-*N*-vinylpyrrolidone or polyvinylalcohol solution followed by a second treatment with an aqueous solution of a protein whose molecular weight was lower than that of the proteins to be separated. Possibly, displacement of the weakly adsorbed coils by the stronger interacting proteins produce an additional shrouding of the polymer-coated supports. After the weakly adsorbed portion was replaced, the stability of the mixed adsorption layer was higher.

The suspension of influenza virus, strain A/X-53, was separated from the constituents of alantoic fluid by preparative size exclusion chromatography on a 10×120 cm column. The hemo-agglutination method revealed an elevated level (93%) of viral activity. The leakage of the bonded phases can be more efficiently minimized, however, with the use of positively charged polymers.

3.2 Adsorption of Positively Charged Polymers

Poly-1-vinyl-1,2,4-triazole (PVT, M = 15000–40000) was found to be an effective surface modifier irreversibly adsorbed on wide porous glasses [32]. Apparently, multiple hydrogen bonding and/or electrostatic interaction of heterocyclic bases with acidic silanols are responsible for the adsorption behavior of the polymer. PVT-covered porous glasses were successfully used for the isolation of influenza, fowl plaque and Sendai viruses from concentrated alantoic fluids as well as for the isolation of bovine adeno- and rotaviruses from the concentrated cultural fluids (see Fig. 1). Purified viruses retained their antigenic and immunogenic activities. The preserved double-layer structure of rotaviruses can be seen using electron microscopy (see Figs. 2A–C). More than 50 chromatographic passages were carried out on the PVT column before regeneration of the packing [33].

Cationic derivatives of dextran were used to synthesize carbohydrate polymeric stationary phases of as support materials for affinity chromatography. After being irreversibly adsorbed on wide porous silica gel, diethylaminoethyl (DEAE) derivatives of dextran formed a stable coating and could be activated by cyanogen bromide for further immobilization of tetanus antigen. Industrial extraction of antitetanus antibodies from placental blood and plasma was carried out with this adsorbent. For regeneration, the packing was separately washed with 2% formaldehyde, 25% ethanol, 0.1 mol/l HCl, 0.1 mol/l NH_3, 4 mol/l NaCl, 3 mol/l NaI and 8 mol/l urea without any loss of capacity or specificity after 10 cycles of chromatography and washings [34].

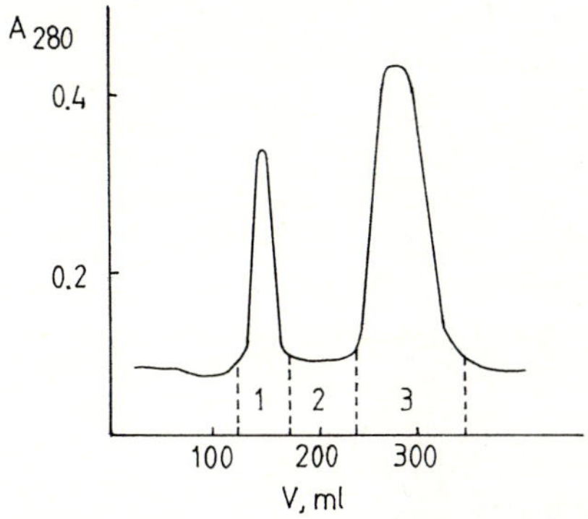

Fig. 1. Preparative separation of the components of the concentrated culture fluid on the PVT-porous glass column. (*1*) fraction of purified rotaviruses, (*2*, *3*) other components of the culture fluid [32]

The adsorption of cationic polymers, therefore, seems to be a promising method for the formation of stable biocompatible bonded phases. Recently, Josefonvicz et al. studied dextran-coated silica packings for high-performance liquid chromatography of proteins [35]. Dextrans with molecular weights of 40000, 70000, 500000 and DEAE-groups content of 3.8–4.7% of carbon were adsorbed on 1250 Å pore size silica gel. The calibration plots log $M - V_{el}$ were found to be linear and very similar for these three composite packings. This indicated that the molecular weight of the DEAE-dextran had no influence on the neutralization of surface silanols. In fact, the number of dextran units bearing DEAE groups was more important for the properties of the stationary phase. Preferably, it had to fit the range 4–10%. Below 4%, the neutralization of negative charges was not completely achieved. In contrast, above 10% an excess of positive charges was present on the coated surface, which interfered with the size exclusion mechanism. Such packings exhibited anion-exchange properties. Among the proteins tested, only cytochrome C was abnormally retained by the modified silica. This effect was probably due to the high pI value and the small size of the protein. Also, it testified to the fact that the dextran coating was not dense enough to prevent contacts of proteins with the silica support completely. Similar effects have already been mentioned for the PEO-coated silicas [29].

In a subsequent paper, the team group described the use of dextran-coated silicas for high-performance affinity chromatography (HPAC) of proteins [36]. Immobilization of protein A, concanavalin A and heparin was carried out on the supports by means of conventional methods of matrix activation by 1,1-carbonyldiimidazol and butandioldiglycidyl ether. The coupling yields of the coated silica supports were similar to those obtained with commercial polysaccharide-based supports. In order to study their performance in HPAC, human IgG, ovalbumin, antithrombin-III (AT-III) and thrombin were eluted from the supports grafted with ligands mentioned above. Recovery of IgG was 84%, of AT-III 78–80%, and of thrombin 85%.

Alpert and Regnier [37] found that polyethyleneimine (PEI, M = 600) was readily adsorbed by various porous silicas from a methanol solution. After the reversibly adsorbed portion of the polymer was washed out, the content of

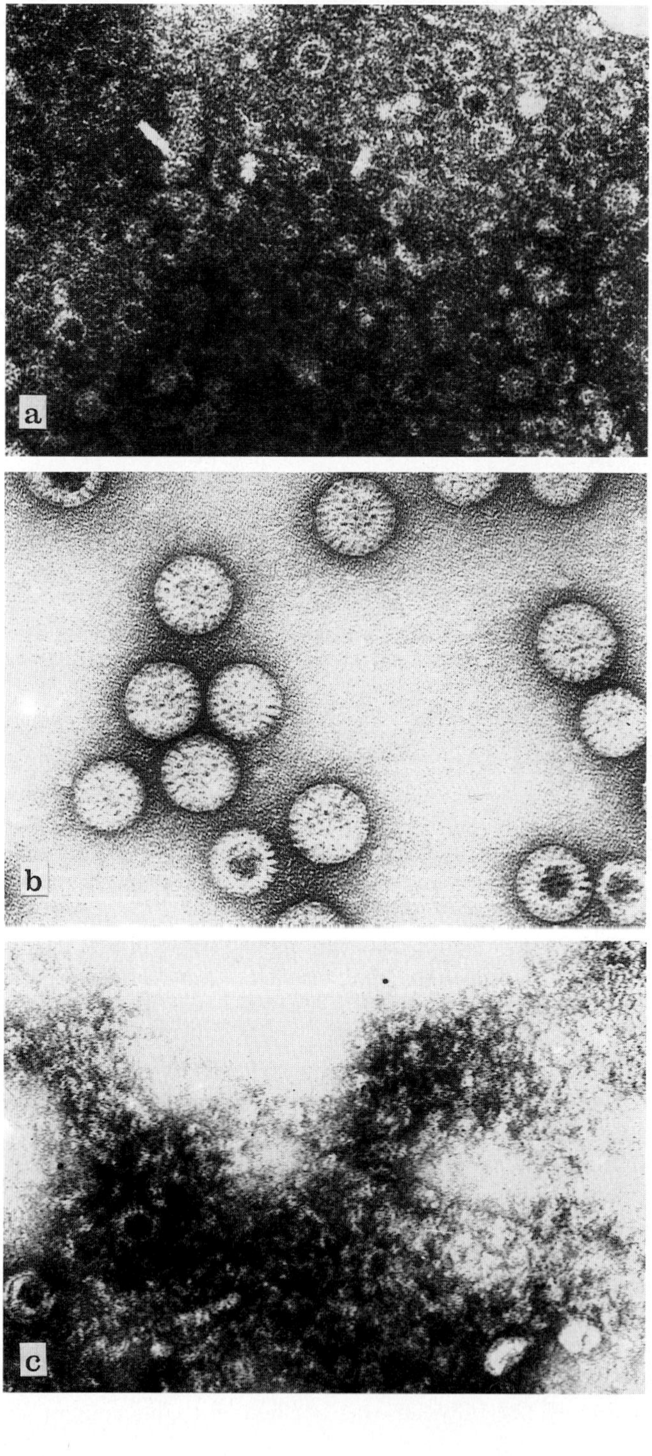

Fig. 2 a–c.

irreversibly bound portion was quantified. Ion-pairing capacities (IPC) of the coated silicas estimated by picric acid titration were found to be proportional to the surface area of silica gel. However, the specific IPC (IPC/m^2) of the 500 Å-support (9.0 μmol/m) was greater than that of the 100–120 Å pore diameter supports (4.9–5.0 μmol/m^2). The prepared PEI-coating could be cross-linked with pentaerythritol tetraglycidyl ether (PETE). Elemental analysis of this coating prepared on LiChrospher Si 500 (10 μm) gave a C/N ratio of 2.89. This ratio corresponds to a model in which each PEI molecule has three of its nitrogen atoms cross-linked to adjacent PEI molecules. Some protein mixtures (human serum, rat kidney homogenate) were resolved on these anion-exchangers by means of a gradient of sodium acetate.

The resolution of these columns for protein mixtures, however, was comparably poor. The peak capacity for human serum albumin was near 3 during 20 min gradient elution. Improvement has been reached by covalent binding of PEI (M = 400–600) onto a 330 Å silica of 5 μm particle size [38]. The peak capacities of ovalbumin and 2a_1-acid glycoprotein were 30–40 ($t_{gradient}$ = 20 min). Enhanced peak capacity and resolution probably were due to the more diffuse structure of PEI coupled to silane moieties than that of strictly adsorbed on silica and cross-linked (see Sect. 2.2). Other applications of covalently adsorbed PEI are discussed in Sect. 4.1.

It is notable that small peptides were retained on a cross-linked PEI column [37] but polypeptides over 20 residues were weakly retained or not at all. Possibly, small peptides were able to penetrate into the depth of the bonded phase while larger peptides could not. These packings were more useful for the separation of oligonucleotides, i.e. more acidic compounds. Lawson et al. [39] have shown the applicability of PEI-silicas for assessing the purity of precursor blocks, monitoring the chemical synthesis and isolating reaction products after synthesis.

Jastrebov et al. prepared PEI-coated silica gels (pore diameter 500 Å) by adsorption of the polymer (M = 30000–40000) from a solution in alcohol followed by cross-linking by dibromoethane and alkylation by methyl iodide. The packings were tested for separation of synthetic oligonucleotides. The elution by potassium phosphate gradients in 30% acetonitryle allowed the isolation of the oligonucleotides with a chain length more than 50 bases from the reaction media [40].

The above results proved the potential viability of the adsorbed hydrophilic macromolecules as bonded phases in chromatography of biopolymers but it must be admitted that additional crosslinking of previously adsorbed macromolecules is usually needed in order to obtain stable composites. The cross-linked bonded polymeric phases, however, may suffer from the restricted flexibility of the chain segment and their steric repellency may be diminished. Moreover, the conformational adaptivity of cross-linked chains for binding with solutes is poorer than that of grafted or chemically bound macromolecules.

◀ **Fig. 2a–c.** Electron micrographs of the chromatographic fractions: (**a**) concentrated culture fluid containing rotaviruses (×30000), (**b**) purified rotaviruses (fraction 1, Fig. 1, ×50000), (**c**) other components of culture fluid (fraction 3, Fig. 1, ×30000) [32]

The most reliable methods of the preparation of stable adsorbents involve, however, a covalent attachment of the polymeric stationary phases to the solid supporting material. In addition, the more diffuse interfaces formed in this case (see Sect. 2.2) are often favourable for the separation of proteins.

4 Chromatographic Packings Prepared by Covalently Bound Polymers and Oligomers

4.1 Coupling via Reactive Groups in the Chains

Macromolecules bearing reactive groups in the repeat units along their chains are capable of multiple interaction with the matrix. As early as 1973, Wilchek prepared Sepharose-based supports chemically modified by chemisorbed polylysine and polyvinylamine [41]. The leakage of dyes covalently bonded to these supports was reduced remarkably as compared to non-modified Sepharose activated by cyanogen bromide. Thus, stable and high capacity affinity adsorbents could be prepared by the introduction of macromolecular spacers between a matrix and a biospecific ligand.

Apart from the mentioned advantages, the polymeric reagents covalently adsorbed by silica also diminish its inherent non-specific adsorptivity. One of the ways to synthesize a polymeric modifier of this type is a copolymerization of a vinylsilane with a compound of the desired functionality. The segments carrying silyl groups will condense with the surface silanols forming “anchors” or “trains”.

Meiller et al. [42] used a silane of the following general formula

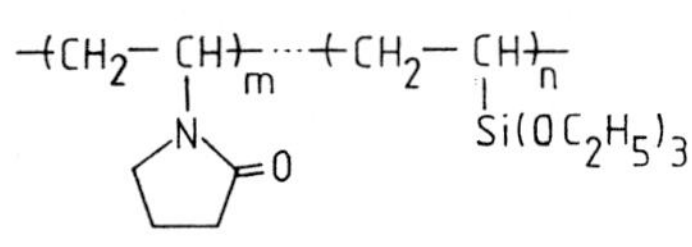

In which the ratio m/n is close to 3. The silane was produced by free radical copolymerization of vinyltriethoxysilane with *N*-vinylpyrrolidone. Its number-average molecular weight evaluated by vapour-phase osmometry was 3500. Porous silica microballs with a mean pore diameter of 225 Å, a specific surface area (S_{sp}) of 130 m^2/g and a pore volume of 0.8 cm^3/g were modified by the silane dissolved in dry toluene. After washings and drying, 0.55% by weight of nitrogen and 4.65% of carbon remained on the microballs. Chromatographic tests carried out with a series of proteins have proved the size-exclusion mechanism of their separation.

The protein recovery was found to be 95% of the amount injected, whereas, on the untreated carrier they were almost totally irreversibly adsorbed. Meanwhile, some reduction in the pore volume of the carrier could be deduced from the results of the chromatographic test. The calculated pore volume available for phtalic acid was 0.67 cm^2/g (V) whereas for cytochrome C – 0.5 cm^2/g. A detailed description of the experiment allows the evaluation of the effective thickness (t_{eff}) of the polymeric stationary phase. The t_{eff} calculated as V/S_{sp} is 2.3 nm. The value obtained is substantially lower than the persistence length of the copolymer that can be evaluated as 5–6 nm. Thus, it becomes evident that the conformation of the chemisorbed macromolecules is not flat and they occupy some volume by their loops and tails. These observations agree with the predictions of the theory (see Sect. 2.2). On the other hand, it is likely that it was the diffuse structure of the bonded phase that provided porous silica with inertness towards proteins and ensured the quantitative recovery of the latter from the packing.

An approach related to the covalent coating of porous silica by polymeric derivatives of vinylsilane has been recently taken by Kurganov et al. [43]. A copolymer of the formula

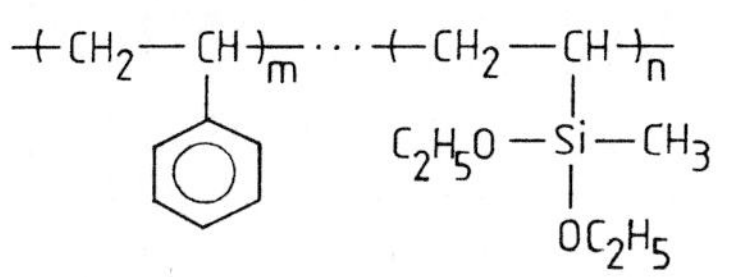

n : m = 12 : 1 and M = 10000 was used for the synthesis of the adsorbents based on porous silicas with pore diameters (d_p) from 6 to 200 nm. The radius of a random polymer coil was calculated by the authors to be 1.4 nm. The surface concentration of the adsorbed copolymer corresponding to the hypothetical monolayer was then evaluated as 7 μmol of monomer units/m^2. The experimental values for Zorbax PSM 1000 and 500 are of the same order, namely 8.4 and 9.3 μmol/m^2, respectively. The values calculated for LiChrospher Si 500 and the spherical silica of d_p = 200 nm are higher than μmol/m^2 (12.3 and 15.5 mol/m^2, respectively). When the mean pore diameter of the starting silica falls into mesopore range ($2 < d_p < 50$ nm), the polymer has limited access to the pores in the immobilization procedure. Thus, for Zorbax PSM 60 and Silasorb Si 600 with a mean pore diameter of 6 and 7.5 nm, respectively, the surface concentration of monomer units are 3.6 μmol/m^2 and 3.5 μmol/m^2.

The effect of coating on the specific pore volume (SPV) and the pore-size distribution (PSD) of silicas was studied by means of mercury porosimetry (MP) and size exclusion chromatography (SEC) of linear polystyrenes. The coating slightly changes the PSD as studied by MP measurements. The pore-size reduction is seen to be more dramatic from SEC measurements: the mean pore diameter at 50% of the cumulative volume is about 80 nm before and about 40 nm after immobilization of the polymer. The discrepancy in PSD between the two methods is due to the fact that the samples were measured using different techniques. In MP, mercury is forced by pressure into the pores of a dry powder. In SEC, the pores are filled with benzene as solvent and the polymer coating is solvated and might swell. This extension of volume leads to pore blockage and limits the access of polystyrene molecules in SEC experiments.

If the swelling effects in the polymeric phase are considered, the evaluation of the adsorbed macromolecules as a "monolayer" seems to be too approximate. The mean-square end-to-end distance of the polystyrene coil (M = 10000) in

Θ-solvent is about 7 nm. Since toluene is a good solvent for this polymer, even larger dimensions of the copolymer coils could be expected for the sample. Thus, the diffuse layer of interpenetrating coils is the better model to explain the enhanced amounts of the adsorbed polymer for the wide-pore silicas.

The described polystyrene-coated silicas were studied as packings for reversed-phase high-performance liquid chromatography of proteins and polypeptides [44]. The relationship between k′ of the tested proteins and acetonitryl content in the isocratic eluents was essentially the same as that for conventional *n*-octyl and *n*-octadecyl reversed-phase packings. The polystyrene-coated RP-materials easily resolved mixtures of peptides and proteins under gradient elution conditions. Peak capacities of about 45–55 were observed for ribonuclease, cytochrome C and lysozyme during 20 min gradient on polystyrene-coated Zorbax 500 packing and 65–75 on analogous Zorbax 1000 packing under the same conditions.

Silica-based cation-exchange adsorbents for chromatography of proteins were prepared by chemical fixation of PEI, M = 600, on Vydac TP 201 silica treated with 2(carbomethoxy)ethyltrichlorosilane (CMETS) [45]. Elemental analysis showed that the product of the reaction between CMETS-silica and PEI contained 3.48% carbon and 1.33% nitrogen. It was calculated that approximately 340 μmol of CMETS and 68 μmol of PEI were bound per gram of silica. After acylation of the residual amines of PEI by glycolic anhydride, approximately 460 μmole carboxyls per gram of packing had been introduced. The mixture of albumin, bacitracin, chymotrypsinogen, cytochrome C and lysozyme was separated on the prepared packing using a 0.01–0.5 mol/l gradient of sodium phosphate, pH 6.0). Recovery of trypsin activity from the column was 91–95%.

Tetraethylenepentamine (TEPA) chemically bound on CMETS-silica was further transformed to weak-hydrophobic stationary phases via its acylation by anhydrides or chlorides of carbonic acids of increasing chain length [46]. Thereby, a series of packings with increasing hydrophobicity was prepared. All the proteins tested, except lysozyme, eluted in the void volume at low ionic strength. High ionic strength was therefore required to adsorb the proteins. Mixture of cytochrome C, conalbumin and β-glucosidase was separated in a 20-min linear gradient from 3.0 mol/l sodium sulphate in 0.1 mol/l potassium phosphate buffer, pH 7.0, to water on 0.41 × 4 cm butyrate column. Acylation of PEI-coatings by mixtures of anhydrides gave a possibility to prepare the packings with appropriate ligand density and selectivity. At high phenyl ligand densities albumin, conalbumin, lactoperoxydase and α-chymotrypsinogen were not eluted from the column during the decreasing salt concentration gradient. In fact, once adsorbed, these proteins required an organic solvent for elution. At the same time, the recovery of enzyme activities after hydrophobic-interaction chromatography on the butyl packing is higher than after reversed phase chromatography of the same enzymes on the SynChropack RP-8 column [46].

The chemical adsorption of a relatively high molecular weight neutral polymer (poly(succinimide), M = 13000) on aminopropyl-Vydac 101 TP silica gel was applied by Alpert [47, 48] to prepare a reactive composite support for use in cation-exchange [47] and hydrophobic-interaction [48] chromatography of proteins. The formation of the bonded polymer layer proceeds through acylation of aminopropyl-silica according to the scheme:

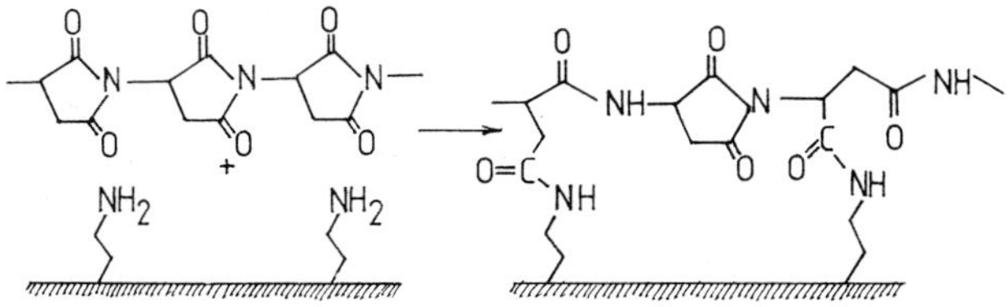

The course of the reaction was followed by measurements of the amino group content carried out with picric acid as a reagent. The polymer was found to acylate 43% of aminopropyls within 0.5 h and 54% within 24 h. These data agree with the findings of Nomura et al. [49] who observed non-quantitative (40–80%) acylation of aminopropyl-silicas by low molecular compounds such as propionyl chloride, benzoyl chloride and stearyl chloride. Furthermore, the conversion of amino groups by poly(succinimide) seems to proceed even faster and more completely than one could expect for a bulky polymeric reagent. Masking of some of the amino groups by the chemisorbed polymer may be the possible reason.

Alpert has shown [47] that poly(succinimide)-silica can be further hydrolyzed to poly(aspartic acid)-silica or condensed with β-alanine in aqueous solution to form a covalently bonded copolymer of 2-carboxyethyl aspartamide and aspartic acid. The content of carboxyl groups generated by this way has not been quantified directly, but the cation-exchange hemoglobin capacity has been measured for a series of the packings. Thus, the optimal concentration of poly(succinimide) used in the synthesis was found to be 2–5%.

Protein mixtures were well resolved on poly(aspartic acid)-silica columns using 0.05 mol/l phosphate buffer, pH 6.0 and a gradient of sodium chloride from 0 to 0.6 mol/l. The columns displayed a high capacity and selectivity. Figure 3 shows the separation of several standard proteins with isoelectric points ranging from 4.7 to over 11. Peaks are sharp and show minimal tailing. The poly(aspartic acid) coating was quite stable: the columns lasted for hundreds of hours of use without decrease in efficiency and capacity.

The high ion-exchange capacity of the coating and its ready release of the adsorbed proteins were ascribed to its hydrophilic, polypeptide nature, and to the location of ionized groups on "branch ends" removed from the solid surface. Such "fuzzy" coatings have a higher surface area, therefore their hemoglobin ion-exchange capacity is also several times higher than that of the carboxymethyl-polyamide (CM-polyamide) material developed by Regnier et al. [45]. Poly-(succinimide)-silica was also assumed to be a suitable support for affinity chromatography or enzyme immobilization.

A series of bonded poly(alkyl aspartamide) coatings was prepared on silica by analogy to the method described above. Poly(succinimide) coating was reacted with *n*-alkyl- and arylalkylamines in dimethylformamide to yield a series of hydrophobic adsorbents. Poly(propyl aspartamide)-silica (PolyPROPYL A) showed the maximal reversible hydrophobic binding of hemoglobin among the C1–C5

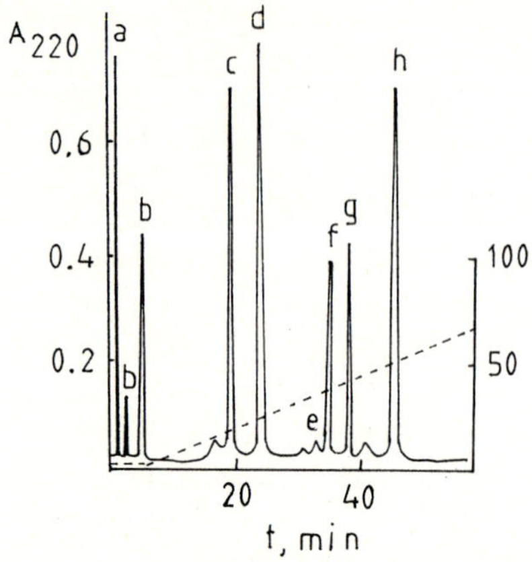

Fig. 3. Cation-exchange chromatography of protein standards. Column: poly(aspartic acid) Vydac (10 μm), 20 × 0.46 cm. Sample: 25 μl containing 12.5 μg of ovalbumin and 25 μg each of the other proteins in the weak buffer. Flow rate: 1 ml/min. Weak buffer: 0.05 mol/l potassium phosphate, pH 6.0. Strong buffer: same +0.6 mol/l sodium chloride: Elution: 80-min linear gradient, 0–100% strong buffer. Peaks: *a* = ovalbumin, *b* = bacitracin, *c* = myoglobin, *d* = chymotrypsinogen A, *e* = cytochrom C (reduced), *f* = ribonuclease A, *g* = cytochrome C (oxidised), *h* = lysozyme. The cytochrome C peaks were identified by oxidation with potassium ferricyanide and reduction with sodium dithionite [47]

packings and thus appeared to be a good column material for general purpose hydrophobic-interaction chromatography of proteins. PolyETHYL A may be useful for more hydrophobic proteins. Furthermore, a complementary selectivity was observed between these two columns. For example, the relative positions of ribonuclease and myoglobin are reversed on PolyPROPYL A and PolyETHYL A columns (see Fig. 4). This is probably due to the fact that myoglobin has a greater proportion of its non-polar residues in the pocket domain than does ribonuclease; the former, therefore, is more sensible to the ligand-type hydrophobic interaction and strongly interacts with C3-packing.

The resolution of the same standard proteins was performed with a SynChropack PROPYL column prepared according to Regnier et al. [46] in order to compare the selectivity of the polyaspartamide-based coatings with that of one based on a different polymer. Retention times are somewhat longer on the SynChropack PROPYL column than on the Poly-PROPYL A column (see Fig. 4). Since the hydrophobic ligand in both coatings has the same nominal length, the effect was ascribed by the author to a greater ligand density on the surface of SynChropack PROPYL. This explanation, however, seems to be inconsistent with the enhanced hemoglobin capacity observed on the poly(aspartic acid)-silica packing as compared with CM-polyamide material of SynChropack type [45, 47].

The most possible reason may be in the higher free energy of the protein adsorption on PolyPROPYL A materials. Chemisorbed neutral poly(succinimide) of molecular weight 13000 apparently forms a diffuse interface as predicted by theory (see Sect. 2.2). Controversially, a short polyethyleneimine exists on a surface in a more flat conformation exhibiting almost no excluded volume and producing

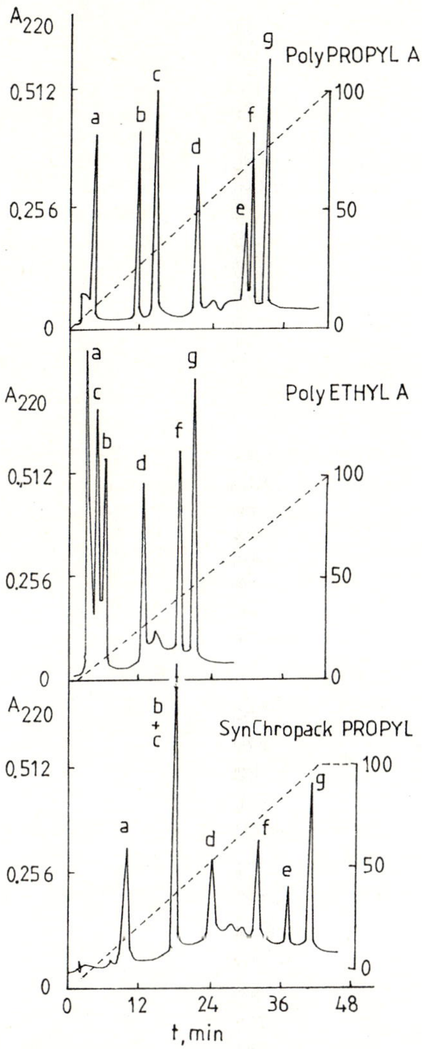

Fig. 4. HPHIC of standard proteins on the weak hydrophobic columns. The SynChropack PROPYL column was 25 × 0.41 cm; Poly(alkyl aspartamid)-silicas were packed into 20 × 0.46 cm columns. Sample: 25 μl containing 25 μg of each protein in buffer A. Buffer A: 1.8 mol/l ammonium sulphate + 0.1 mol/l potassium phosphate, pH 7.0. Buffer B: 0.1 mol/l potassium phosphate, pH 7.0. Gradient: 40-min linear 0–100% buffer B. Flow rate 1 ml/min. Detection: A_{220} = 1.28 a.u.f.s. Peaks: *a* = cytochrome C, *b* = ribonuclease A, *c* = myoglobin, *d* = conalbumin, *e* = neochymotrypsin, *f* = α-chymotrypsin, *g* = α-chymotrypsinogen A [48]

no entropic repulsion towards adsorbing proteins. As a result, retardation of the latter occurs stronger on SynChropack packing.

Several wide-porous affinity and size-exclusion chromatographic supports were prepared by Ivanov, Zubov et al. by means of acylation of aminopropyl-glass supports by copolymers of *N*-vinylpyrrolidone (N-VP,1) and acryloyl chloride (AC,2), M = 7700 and 35000 respectively [50, 51]. The copolymers prepared by free radical copolymerization contain their units almost in equimolar proportion, with high tendency to alternation expected from the copolymerization parameters (r_1 = 0.035, r_2 = 0.15 [52]). Residual carbonyl chloride groups of the chemisorbed copolymer could be transformed to 2-hydroxyethylamides which were solely

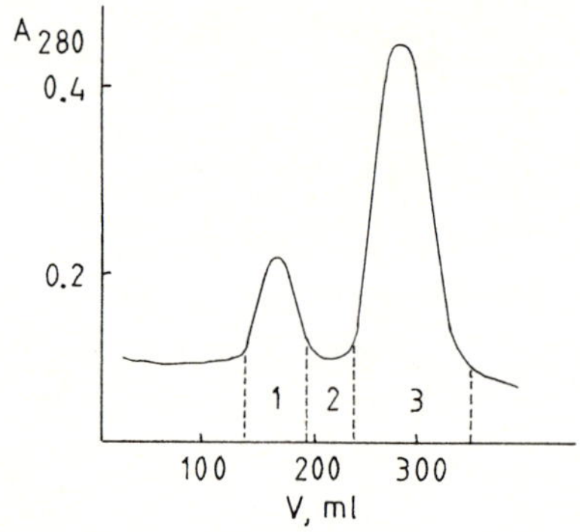

Fig. 5. Preparative separation of the components of concentrated culture fluid on the porous glass chemically modified by the copolymer of N-VP and N-HEAA. 30 ml of concentrated culture fluid was applied to the column (2.3 × 90 cm) equilibrated with 0.01 mol/l phosphate buffer, pH 7.4 and eluted with this buffer at flow rate 300 ml/h. (*1*) – fraction of purified rotavirus, (*2, 3*) – other components of the culture fluid [51]

formed in the reaction with ethanolamine [51]. Also, the affinity ligands bearing primary aminogroups could be coupled to the reactive composite carrier [50].

Chemically attached copolymers of *N*-vinylpyrrolidone (N-VP) and *N*-(2-hydroxyethyl)acrylamide (N-HEAA) steeply decrease the inherent glass adsorptivity which is observed for proteins in aqueous buffer solutions. Thus, it became possible to apply the prepared materials to the size exclusion chromatography of viruses and ribosomes.

Influenza, fowl plaque or Sendai viruses as well as bovine rotaviruses and adenoviruses were isolated thereby from the concentrated alantoic and cultural fluids in preparative scale. Typical separation profile is shown in Fig. 5 and consists of two peaks corresponding to the viral suspension (1) and admixtures (2,3). The main structural proteins of the purified virus are visualized by SDS PAGE electrophoresis as shown in Fig. 6. In most cases a negligible or a very slight decrease of antigenic or agglutination activity was observed for the purified virus. It testifies to the high inertness and good shielding effect produced by the copolymer coating. As compared to PEO or PVP coatings (see Sect. 3.1) it is much more stable and could work for several years. From time to time, a treatment of the packing by 50% aqueous *n*-propanol solution (incubation for 2–3 h at room temperature and periodical mechanical stirring) was used to remove any remained lipids. This procedure did not diminish the quality of separation on the repacked column.

Indeed, the polymeric interface seems to be highly diffuse and hydrophilic because copolymers of *N*-vinylpyrrolidone and *N*-(2-hydroxyethyl) acrylamide are readily soluble in water [53]. Besides, aminopropyl-glass adsorbs the acryloyl chloride copolymer so that only 10% of its active functions become amidated. The rest is located on the loops and tails of the attached macromolecules [51]. Thus the steric repulsion of the bonded phase is a probable reason for the high inertness of the packing towards viruses.

In addition to viruses, ribosomes 70S were separated from RNA on the column packed with the same support. The yield of both components was quantitative.

Porous silicas coated with N-VP – AC copolymer may be used as an activated support for the immobilization of the biospecific ligands [50] or for the synthesis of hydrophobic adsorbents [54].

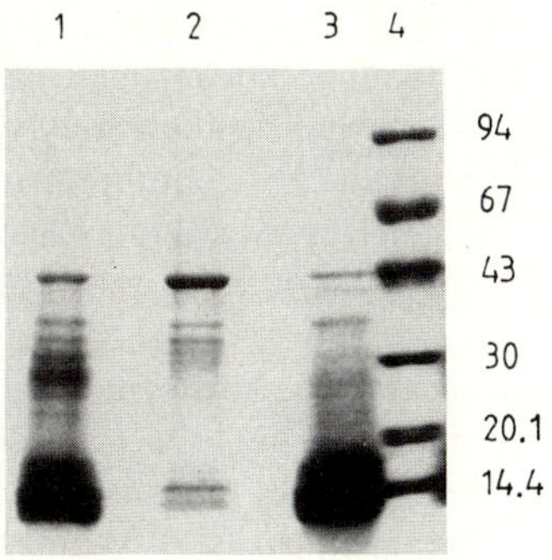

Fig. 6. SDS-PAGE of chromatographically purified rotavirus: (*1*) concentrated culture fluid containing rotavirus before chromatography (Fig. 2, A), (*2*) purified rotavirus (fraction 1, Fig. 5), (*3*) other components of the culture fluid (fraction 3, Fig. 5) [51]

These carbonyl chloride-activated carriers are unstable on storage, so ligand coupling to them has to be undertaken immediately after chemisorption of the copolymer. To overcome this drawback, more stable activated carriers have been synthesized by treating aminopropyl-silicas with poly(*p*-nitrophenyl acrylate) and acetic anhydride.

As revealed by IR-spectroscopy, the attachment of the polymer proceeds via acylation of aminopropyls: absorbances of both amides (1650 cm^{-1}) and esters (1740 cm^{-1}) contribute to the spectrum of polyacrylate-coated aminopropyl-Aerosil (specific surface area 175 m^2/g) [55]. During the reaction, the accumulation of *p*-nitrophenyl ester groups in the support is accompanied by the liberation of *p*-nitrophenol into the contacting solution. Thus, the evaluation of the conformational state of adsorbing macromolecules can be performed by the simultaneous study of both processes by UV-spectroscopy as shown in Fig. 7. Apparently, at

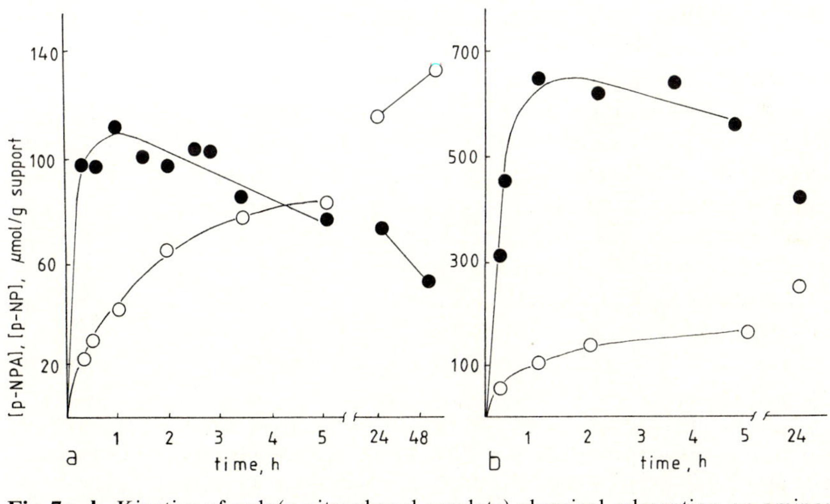

Fig. 7a, b. Kinetics of poly(*p*-nitrophenyl acrylate) chemical adsorption on aminopropyl-Aerosil at 25 °C in dimethylsulphoxide. *Filled circles:* ester group content (µmol/g support), *empty circles: p*-nitrophenol release (µmol/g support), **a** – 1% solution; **b** – 5% solution [55]

first, a macromolecule interacts with the reactive support via a few monomer units. Then, the relaxation of primarily bound coils allows them to adopt a flatter conformation. However, the addition of excess acetic anhydride into the reaction media may restrict this process to the appropriate value: up to 70% of aminopropyls become acetylated under these circumstances [49]. In this way, ionizable primary amines are transformed to neutral acetyl amides, diminishing possible nonspecific adsorptivity of the composite carrier.

Reactive *p*-nitrophenyl ester groups of the carrier can be transformed to *N*-alkylamides by the reaction with primary alkylamines. Thus, two weak hydrophobic adsorbents were prepared. Both were based on wide-porous glass (d_p = 2000 Å) with their surface being derivatized by chemisorbed *N*-butyl polyacrylamide (Butyl-PG) or *N*-benzyl polyacrylamide (Benzyl-PG). Studies of adsorption isotherms and of desorption conditions for model proteins (ovalbumin and γ-globulin) revealed that Butyl-PG resembles semirigid synthetic adsorbent Butyl-Toyopearl 650 C (TOSOH, Japan), but had a 0.6-fold capacity and a 2.8-fold permeability of the latter for γ-globulin [56]. Several kinds of adsorbents for hydrophobic interaction chromatography (Butyl-PG, Benzyl-PG, Butyl-Toyopearl 650 C, Phenyl-Sepharose CL-4B) were compared in a model purification of high-molecular weight (300000–400000) surface antigens of human and avian adenoviruses (hexon antigens). All the adsorbents tested revealed similar selectivities for the viral proteins, Butyl-PG having the maximal capacity for the hexon egg-drop syndrome-76 virus (EDS-76). The chromatography of crude alantoic fluid from duck embryos infected with EDS-76 gave rise to 4-fold concentration of EDS-76 hexon with simultaneous 20-fold purification (see Figs. 8 and 9). The same degree of purification and 10-fold concentration were obtained during chromatography of crude cultural fluid containing hexon of human adenovirus serotype 5.

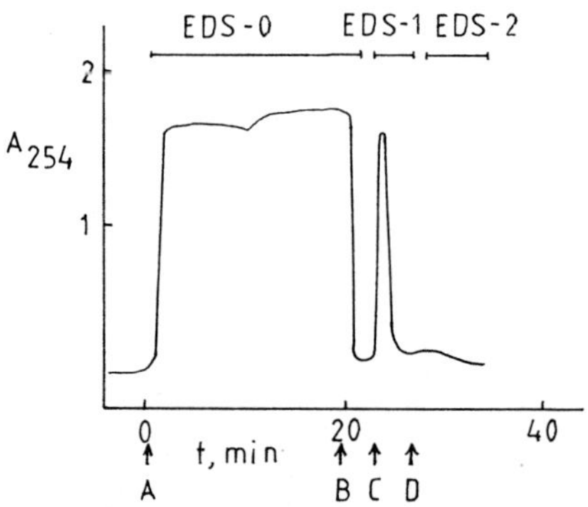

Fig. 8. Preparative isolation of hexon antigen of EDS-76 by hydrophobic-interaction chromatography on Butyl-PG column (2 × 5 cm): (*A*) application of the allantoic fluid diluted (1 : 5) by 50 mM potassium acetate, pH 4, 130 ml; (*B*) 0.01 mol/l potassium acetate, pH 5.5; (*C*) 0.01 mol/l potassium bicarbonate pH 8.0, 10% isopropanol; (*D*) 0.01 mol/l potassium carbonate pH 9.6, 10% isopropanol. *EDS-0* – components of alantoic fluid eluted with buffer A, *EDS-1* – desorbed hexon fraction eluted with buffer C, *EDS-2* – fraction desorbed with buffer D [56]

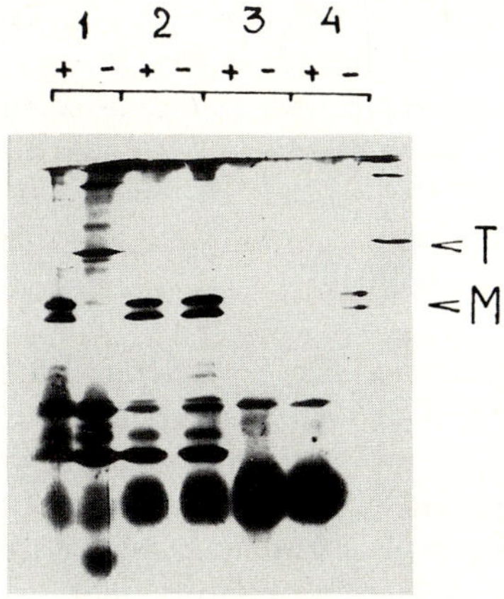

Fig. 9. SDS-PAGE of chromatographically purified hexon antigen (EDS-76): (*1*) allantoic fluid before chromatography; (*2*) allantoic fluid acidified to pH 4; (*3*) fraction eluted by 0.01 mol/l potassium acetate, pH 5.5 (Fig. 8, A); (*4*) fraction of hexon antigen eluted by 0.01 mol/l potassium bicarbonate, pH 8.0, 10% isopropanol (Fig. 8, C), "Plus" denotes boiling in the dissociating buffer solution, "minus" – incubation in the same buffer at 25 °C. *Arrows* indicate the bands of monomer (*M*) and trimer (*T*) of the hexan antigen [56]

N-Butyl polyacrylamide bonded phase was also synthesized on the 500 Å-pore fine "Armsorb" silica gel (YERONEM, Yerevan, USSR) to prepare a weak hydrophobic adsorbent for high-performance chromatography of proteins (Fig. 10). The packing exhibited a high peak capacity and selectivity for small proteins when used in the reversed-phase chromatographic mode. Positions of eluted proteins indicated a weaker retention as compared with Nucleosil RP Protein 300-7 column (Macherey Nagel, Germany). Apparently, this was a consequence of enhanced hydrophilicity of the polyacrylamide coating. Meanwhile, the recombinant insulin sample could be resolved better with the use of *N*-butyl polyacrylamide column. No change in the elution behavior of proteins was marked over one year: the column was currently used for the determination of recombinant proinsulin and insulin in culture fluids. Thus, reasonable stability of the packing has been proved [57].

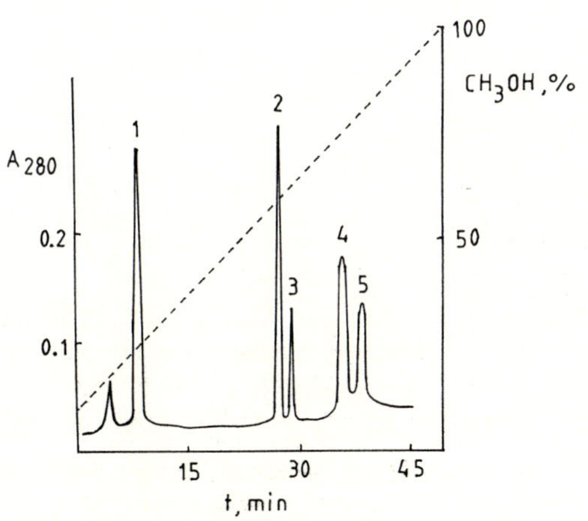

Fig. 10. HPLC of proteins (commercial samples) on the *N*-butyl polyacrylamide coated silica gel column. Sample: 20 μl of 5–15 mg/ml protein solution in buffer A. Buffer A: 10% methanol, 0.2 mol/l ammonium acetate, pH 4.5. Buffer B: methanol. Gradient: 50-min linear, 0–100% B. Flow rate 0.8 ml/min. Peaks: (*1*) – lysozym, (*2, 3*) – insulin, (*4, 5*) – myoglobin [57]

Poly(*p*-nitrophenyl acrylate)-coated wide-pore glass (WPG) was also used as an activated carrier for the immobilization of biospecific ligands and enzymes. A detailed description of properties of these sorbents and catalysts as well as some specific features of their functioning is given in Sect. 6.

4.2 End Group Coupled Chains: Polyalkyleneoxides

End-group grafting of a polymeric or oligomeric modifier is preferable for polyethylene oxide and polypropylene oxides. As early as 1974, Meiller et al. [42] prepared a reactive silane by means of hydrosilylation reaction between oligoethyleneoxide allyl ether and dimethylchlorosilane. Silica gel microballs of mean pore diameter 225 Å, pore volume 0.8 ml/g and specific surface area 130 m^2/g were treated with this silane in toluene to put oligoethyleneglycol grafts onto the surface of the carrier. The inner pore volume of the modified microballs was 0.75 cm^3/g as determined by aqueous size-exclusion chromatography of blue dextran ($M = 2 \times 10^6$) and phthalic acid molecular weight markers. Several proteins (cytochrome C, myoglobin, hemoglobin, γ-globulin and others) were chromatographed on the prepared packing at pH 7.5. In contrast to untreated microballs, virtually complete elution of all proteins has been observed from the modified silica. The positions of these eluates did not exactly obey, however, a size-exclusion mechanism.

In the middle of the 1980s in a number of papers, PEO-silicas were studied as adsorbents for high-performance hydrophobic interation chromatography of proteins and nucleic acids. Miller et al. [58] prepared a series of silanes bearing an alkyl group on the one end of the diethylene glycol moiety and a triethoxysilyl group on the other. Vydac silica gel (7 μm, pore diameter 300 Å) was treated by the prepared silanes so that a series of adsorbents with various hydrophobicity was formed. A mixture of standard proteins was effectively separated on these packings during a descending gradient of ammonium sulphate (3.0 mol/l – 0 mol/l) in 0.5 mol/l ammonium acetate, pH 6.0. The yields of proteins were higher than 90%, whereas recovery of lysozyme enzymatic activity was 170% (summed up from 2 applications).

Chang, El Rassi and Horvath [59] studied the silica bound polyethylene glycol ($M = 400$) as a stationary phase for the same type of chromatography. The method of synthesis included a treatment of 5 μm Hypersil (pore diameter 300 Å) by 3-glycidoxypropyltrimethoxy silane and Carbowax PEG 400. A mixture of six proteins was separated during linear gradient elution (20 min) with decreasing concentration of ammonium sulphate. As far as the proteins could be eluted in the isocratic mode too, the effect of salt concentration (m) in the eluent on the retention of proteins was investigated. Plots of log k′ vs m were linear. The recovery of enzymatic activity of adenosine deaminase obtained during gradient elution was 90%, thus the novel stationary phase could be considered as "soft".

The same research group proved the applicability of PEO-silicas to the separation of ribonucleic acids and studied how the log k′ vs m slopes are affected by the molecular weight of polyethylene oxide, the type of salt used in eluent

(Na_2HPO_4 or $(NH_4)_2SO_4$) and by the silane sublayer structure [60]. The limiting slope of log k′ vs m plots was termed as the hydrophobic interaction parameter (HIP) depending on the polymer-protein contact area. For polyethyleneoxides of molecular weight 400 – 4000, ovalbumin and lysozyme used as sorbates, HIPs were found to increase with the chain length of the stationary phase ligates. This demonstrates the increase of the hydrophobic contact area upon increasing the length of the bound polyether chains.

On the other hand, the HIP value for ribonuclease was practically independent of the length of the polyether ligate. These stationary phases were also employed for the separation of oligophenylalanines containing up to 4 residues by isocratic elution with 0.5 mol/l phosphate buffer, pH 6.3. The retention increments of the Phe residues did not depend on the ligate length, too, and were 0.82 and 0.89 for the stationary phases composed of PEOs (1500 and 4000, respectively).

It is likely that small molecules such as short oligopeptides have almost equal access to the hydrocarbonaceous sublayer at the surface of the bonded phases and thus their retention behavior is not affected significantly by the size of the hydrophilic polyether moieties.

On the other hand, the proteins are believed to interact only with the bound polyether moiety which effectively shrouds the surface region of the stationary phase for bulky protein molecules.

Owing to the weak hydrophobicity of the PEO stationary phases and reversibility of the protein adsorption, some advantages of these columns could be expected for the isolation of labile and high-molecular weight biopolymers. Miller et al. [61] found that labile mitochondrial matrix enzymes – ornitine transcarbomoylase and carbomoyl phosphate synthetase (M = 165 kDa) could be efficiently isolated by means of hydrophobic interaction chromatography from the crude extract.

5 Interfaces Formed by Graft Polymerization

5.1 Procedures of Surface Activation and Grafting

Polymer chains may be attached to the surface of solids not only by binding of preformed polymers but also by graft polymerization. Thus, grafting may involve the growth of polymer chains from active sites of the surface by a step chain polymerization mechanism, as well as chemical bonding of "living" polymer chains to the support surface. However, in the latter case, since polymer molecules must diffuse to the solid surface of a porous material, diffusional limitations and steric hindrances may severely reduce the degree of surface coverage and grafting yield [62]. In contrast, in graft polymerization, diffusion limitations and steric hindrances for both porous and nonporous supports are diminished due to the small size of monomer molecules. Thus, a higher surface concentration and more uniform surface coverage by a grafted polymer layer are possible.

Graft polymerization can be accomplished provided that active sites are available on the surface for reaction with monomers. Two fundamental methods of creating such active sites are known. The first one is formation of active sites directly on the surface of silica substrates with a surface hydroxyl concentration of approximately 4.6 groups/nm^2 [63]. The other approach consists of preliminary chemical modification of the surface hydroxyl groups in order to create new active sites on the surface. These sites on the silica surface can be produced by the formation of siloxane bridges, e.g. utilizing surface coupling agents [64, 65]. Organosilanes are used extensively as coupling agents with the silica and modifiers of interfacial properties. A particularly effective class of silane coupling agents are those of the form $RSiX_3$, where X is hydrolyzable group (i.e. halogen, amine, alkoxy, etc.) and R is a nonhydrolyzable organic group. The desired characteristics of the silane are determined by the choice of the functionality of the R group. For example, coupled vinylsilane provides the vinyl groups, which can react with various vinyl monomers through a free-radical polymerization reaction scheme to form surface-grafted polymers.

The creation of active sites as well as the graft polymerization of monomers may be carried out by using radiation procedures or free-radical initiators. This review is not devoted to the consideration of polymerization mechanisms on the surfaces of porous solids. Such information is presented in a number of excellent reviews [66–68]. However, it is necessary to focus attention on those peculiarities of polymerization that result in the formation of chromatographic sorbents. In spite of numerous publications devoted to problems of composite materials produced by means of polymerization techniques, articles concerning chromatographic sorbents are scarce. As mentioned above, there are two principle processes of sorbent preparation by graft polymerization: radiation-induced polymerization or polymerization by radical initiators. We will also pay attention to advantages and deficiencies of the methods.

Radiation-induced graft polymerization may be carried out by the following procedures [69]:

1) by direct irradiation of the system: support(silica)-monomer. In this case monomers may be in the liquid state (pure or in solution) as well as in the gas or vapor states;
2) by grafting on the support previously irradiated in an inert atmosphere or vacuum ("post-polymerization"). In this case monomers are usually in the gas or vapor state.

Radiation-induced polymerization in the presence of solid supports is realized in the monomer layers adsorbed on the surface of silica. It is assumed that surface silanolic groups are responsible for the polymerization and may give surface radicals $SiO^{\bullet}$ and $Si^{\bullet}$ under irradiation [70, 71]. Using this method, Kosaka et al. [72] polymerized such monomers as styrene, methyl methacrylate, acrylonitrile and allylglycidyl ether on the surface of silica with high yields of grafted polymers (15% to 42%) and highly efficient grafting.

However, the polymerization of monomers in the gaseous phase by first procedure (1) mentioned above leads to the formation of considerable amounts of homopolymers. It is caused by low molecular weight radicals $H^{\bullet}$ and $OH^{\bullet}$, not bounded chemically with the solid support [70, 73]. In this case the subsequent treatment by crosslinking agents or preliminary modification of silica with unsaturated organosilanes are obviously necessary.

Currently, graft post-polymerization of monomers in the gaseous phase (2) is the more widely used process because it has at least two basic advantages. First, side processes of homopolymerization are minimized which reduces the consumption of monomers and makes unnecessary additional treatment of the modified materials with solvents. Second, this method is universal and allows the grafting to the surfaces (such as silica) to be carried out with low radiation yields of active sites as compared to the monomers.

The post-radiation graft polymerization of methyl methacrylate and styrene under normal temperature on irradiated unmodified silica did not permit the attainment of a high degree and efficiency of grafting [74]. As a result of the treatment of silica by unsaturated organosilanes, such as divinyldichlorosilanes, the concentration of active centers formed under the action of irradiation is sharply increased; the same holds true for the yield of grafted polymer and the efficiency of grafting. The degree of grafting increased with increasing absorbed dose and approached a limiting value at high dose [73, 75].

The yield of grafted polymers and the efficiency of grafting may be increased by means of low temperature post-polymerization of monomers. Tetrafluoroethylene was polymerized on the surface of silica with high yields of grafted polymer (up to 30%) [76]. However, a high degree of grafting does not always result in an effective polymeric cover of silica surface. It is determined by specific condition of the polymerization process. For instance, Saburov et al. [77] found that the porous structure of the original silica had been preserved.

The polymerization of vinyl monomers on the surface of silica can be induced also by free radical initiators such as azo-bis-isobutyronitrile (AIBN), di-tert-butylperoxide, benzoyl peroxide etc. The selection of initiator type and method of its introduction in polymerizable systems are determined by the nature of monomers and tasks of investigations. Usually, the following procedures are used:

1) the binding of initiator to the silica surface by means of covalent or other chemical bonds;
2) applying the initiator to the silica surface by means of preliminary physical adsorption from the solution;
3) introducing the initiator into the monomer or its solution.

For chromatographic sorbents it is necessary that the polymeric cover be uniformly distributed over the silica surface and chemically coupled to it. The appropriate introduction of the initiator is one of the decisive steps of this task. The first method (binding to the surface) increases the yield of grafted polymer. However in this case a large amount of homopolymer is formed. This disadvantage could be prevented by the application of hydroperoxide initiators in combination with the proper redox-agents [78–81].

Advanced methods of producing sorbents with a uniform and resistant polymeric cover involve additional steps. The latter may include:

1) using monomers with reactive groups. Kalal et al. [82–84] described the process of polymerization for 2,3-epoxypropyl methacrylate (2,3-EPMA) on porous

glass. The procedure was carried out in two stages, namely, by binding 2,3-EPMA on the glass, and then by polymerization of 2,3-EPMA in the presence of glass modified with bound monomers prepared in the 1st stage. The polymerization conditions were chosen so that the basic parameters of a porous structure remain preserved and polymer layer was distributed over the internal surface of the porous glass uniformly.

2) preliminary activation of the silica surface by unsaturated organo-silica compounds. Silica compounds modified in this way are capable of copolymerizing with vinyl monomers, such as methyl methacrylate [85], acrylonitrile [81, 86], acrylic and methacrylic acid [87], styrene [81, 85, 88], *N*-vinylpyrrolidone [87, 89, 90], etc. Hayakawa et al. [91] used triethoxyvinylsilane-styrene and triethoxyvinylsilane-methylmethacrylate binary monomer mixtures for the graft polymerization initiated by benzoyl peroxide onto various silicates.

3) using crosslinking agents. Polymerization of vinyl monomers in the presence of bifunctional crosslinking agents and suspended silica is the method of sorbent production in a way similar to the polymer adsorption technique. The polymer formed in the solution is adsorbed on the silica surface and is further crosslinked there. The structure of the adsorption layer consists of branched macromolecules and microscopic three-dimensional particles of crosslinked gel. The properties of macroporous silicas modified with copolymers of *N*-vinylpoyrrolidone and dimethacrylates of ethylene- and triethylene glycol by this procedure were investigated [92]. In the following paper a procedure for chemical binding of polystyrene was described [93]. In this process the styrene monomer is attached to the surface in the presence of *t*-butyl peroxide which forms a polymer radical. Further polymerization of styrene occurs with the addition of crosslinking agent (e.g. divinylbenzene) and free radical initiator AIBN.

The above results prove the potential of the graft polymerization technique for the preparation of composite sorbents. The next section will be devoted to the application of such materials in the chromatography of biopolymers.

5.2 Chromatographic Packings Prepared by Graft Polymerization of Monomers

Macroporous silica modified with crosslinked copolymers of *N*-vinylpyrrolidone and dimethacrylates of ethylene- and triethylene glycol were used for the purification by size exclusion chromatography of viruses from admixtures of proteins with high yield [92]. It was indicated that the polymer coating was distributed on the glass surface snugly and uniformly. From this supposition and elemental analysis data the average thickness of the polymer layer was estimated to be about 15–20 Å. However, in the case of porous glass with a pore diameter 20 nm, the investigation of the sorbents by mercury porosimetry showed that porous structure of the original support had suffered essential changes after the polymer grafting. But wider porous samples (with pore diameters up to 200 nm) had virtually identical parameters of porosity before and after modification. Related method of sorbent preparation consists in γ-irradiation of mixture of zeolite and aqueous solution of hydroxyethylmethacrylate (11%) and *N,N'*-methylenebisacrylamide (0,15%) [33]. After immobilization of pigment Cibacron Blue F-3A the sorbent acquired the ability to adsorb serum albumin from solution.

Silica modified with γ-glycidoxypropyl triethoxysilane was used for graft polymerization of various oxirane monomers in presence of $BF_3 \cdot OEt_2$ (0.2–0.5%) [95]. After this procedure the adsorbents were treated by diethylamine, diethylaminoethanol and polyethylenimine in order to obtain ion-exchangers for HPLC of proteins. By means of oxidation by periodate, permanganate and bisulfite in aqueous solution, the grafted copolymers were transformed to carboxylic or sulfuric derivatives. Applications of these adsorbents included chromatographic separation of proteins from blood serum, or components of commercial pancreatine and trypsine inhibitor. Gorkovenko et al. [96] used an analogous procedure for polymerization of levoglucozane and its polymeric analog. Sorbents prepared by this method were utilized for size-exclusion chromatography of proteins with high yields (94–98%).

A series of exclusion and affinity sorbents with stationary phases of copolymerized vinyl monomers containing amide group was created on the base of wide pore silica (p_d = 25–70 nm) chemically modified with γ-glycidoxypropyl trimethoxysilane [97]. *N*-methylolacrylamide was copolymerized with *N*-substituted acrylamides, acrylic acid (activated, for instance, with *N*-hydroxysuccinimide), allylamine or *N*-allylbromacetamide. By using bifunctional vinyl monomer such as *N*-methylolacrylamide the polymerization is, in the presence of a trace of acids or under heating accompanied with additional crosslinking due to the reaction of the *N*-methylol hydroxyl groups with γ-glycidoxypropyl siloxy groups of the inorganic support material or with themself. As a result the polymer is crosslinked as well as covalently bonded to the silanized material.

For the use of such materials in affinity chromatography, the reactive groups have to be coupled to the compounds which on the one end have a corresponding reactive group and on the other an affinity ligand. The surface containing about 100 μmoles of carboxyl groups/g activated with *N*-hydroxysuccinimide were reacted with *p*-aminobenzamidine hydrochloride and were used for trypsin and chymotrypsin isolation. The column was equilibrated with 0.1 mol/l sodium acetate buffer (pH 5.5) containing 0.1 mol/l NaCl and 0.001 mol/l $CaCl_2$. Flow rate was 15.6 ml/h. The column was eluted with 0.1 mol/l glycine-HCl buffer (pH 2.0). It was found that the trypsin so obtained in eluate was purified 7 times and chymotrypsin about 3 times [97].

During the last decade, the composite spheric material TSK-GEL SW developed and produced by "Toyo Soda Manufacturing" was one of the most favorite adsorbents for HPLC of biopolymers. The approach of the manufacturers and the exact composition of the polymeric coating are unknown. However, the coating was apparently constituted by polymers containing *cis*-diol groups, because in article [98] TSK-GEL 4000 SW was used for immobilization of cephalosporin C, linked to the surface by means of cyanogen bromide activation. The material was effectively used for the high performance liquid affinity chromatographic technique of homogeneous purification of active β-lactamase.

The elution of proteins chromatographed on TSK-GEL SW in 0.07 mol/l phosphate buffer (pH 6.8) in the presence of 0.1 mol/l NaCl takes place by gel permeation chromatography mechanism. A mixture of four commercial proteins, albumin (human serum), β-lactoglobulin, myoglobin and cytochrome C, was studied using TSK-GEL 2000 SW column. Human serum and egg white was examined using TSK-GEL 3000 SW [99]. Small deviations from linear calibration (log M vs V_{el}) for some types of proteins were analyzed by Wehr [100]. The weak anion exchangers were prepared by introducing diethylaminoethyl groups into chemically bonded hydrophilic layers of TSK-GEL SW, which is a silica-based support of particle diameter 10 μm and pore diameter 250 Å [101]. Commercial ovalbumin was used as a standard for high performance ion exchange chromatography on this sorbent and linear gradient elution with NaCl to investigate the effects of gradient, flow rate, column length and sample loading on resolution and retention.

When a porous chromatographic support consists of a basic material on which vinyl polymer chains may be grafted in sufficient lateral density, ion exchangers of the following types may be formed from the monomers listed:

$CH_2{=}CHCONH(CH_2)_2N(CH_3)_2$	– "DMAE" type, weak anion exchanger,
$CH_2{=}CHCONH(CH_2)_2N(C_2H_5)_2$	– "DEAE" type, weak anion exchanger,
$CH_2{=}CHCONH(CH_2)_2N^+(CH_3)_3$	– "TMAE" type, strong anion exchanger,
$CH_2{=}CHCOO^-$	– "COO^-" type, weak cation exchanger,
$CH_2{=}CHCONC(CH_3)_2CH_2SO_3^-$	– "SO_3^-" type, strong cation exchanger.

The grafting reaction requires primary or secondary aliphatic hydroxyl functions as initiation sites and uses Ce^{4+} ions as the catalyst. This type of grafting was first developed by Mino and Kaizerman and applied to preparation of the polyacrylamide-polyvinylalcohol copolymer [102].

It is obvious that the density of the hydroxyl groups on the support surface and amount of catalyst used determine the density of grafting. Similarly, the amount of monomer governs the mean length of the grafted chains. Indirect figures and microscopic studies indicate that chains consisting of a few monomers only are equally accessible as long chains, forming brush-like layers up to 50 nm in thickness. The chain length optimally suited for ion exchange process will definitely depend on the type of analyte used. According to present experience, chains of 5–50 monomers seem to function properly for proteins and nucleic acids. The fact that 50 monomers correspond to a length of about 10 nm in the fully extended state emphasizes the importance of using supports with sufficiently wide pores; materials with 100–500 nm pore diameter have been found to be adequate.

Mueller [103–104] has adopted the poly(ethylene glycol)-dextran partition system pioneered by Albertson [105] to develop an efficient chromatographic method for the resolution of proteins and nucleic acids. He used supports such as Fractogel TSK and LiChrospher Diol (Merck) and grafted a polyacrylamide layer at the surface in order to immobilize the dextran rich phase. The loaded support was then equilibrated with poly(ethylene glycol)-rich phase as the mobile phase. The biopolymers distributed themselves between the two phases according to their size, nature of the surface, and net charge. Retention and selectivity were controlled by the type of salt, the salt concentration, the pH, the temperature and the molecular weight of the two constituents poly(ethylene glycol) and dextran. Chromatographic separation was carried out in the gradient elution mode on columns of about 200 × 10 mm with a flow rate of <0.5 ml/min [103]. Column capacity was dependent on the molecular weight of analytes and varied between 2 mg/cm^2 of column cross-section (200 mm length) for 80 kDa protein, and 0.5 mg/cm^2 for DNA fragments of 1300 to 2000 kDa [104].

The separation of some proteins (bovine serum albumin, lysozyme, haemoglobin, chymotrypsinogen A, cytochrome C) and large and small restriction fragments of DNA has been demonstrated by means of ion exchangers prepared from Fractogel and LiChrospher in the conventional way and the polymer-grafting process ("tentacle"-type modification) described above has been demonstrated. Comparative data for the capacities of both types of ion exchangers indicate that a substantial increase in capacity was achieved by tentacle type modification (more than in 2 times).

In Fig. 11, separations of chymotrypsinogen A, cytochrome c and lysozyme on strong cation exchangers carrying SO_3^- groups by gradient elution are shown.

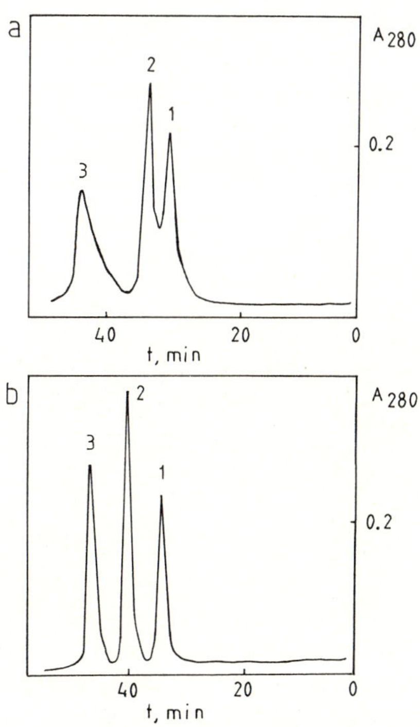

Fig. 11a. Fractionation of (*1*) chymotrypsinogen A, (*2*) cytochrome C and (*3*) lysozyme on strong cation exchangers. *a*) Support: Fractogel TSK 650(s)SP (conventional type); sample, 1 mg each; flow rate, 1 ml/min; column size, 150 × 10 mm I.D. Solvent A = 0.02 mol/l phosphate, pH 6.0; solvent B = A + 1 mol/l NaCl; gradient, 0–10 min, 0% B; 10–70 min, 0–100% B. **b)** Support: Fractogel EMD 650(s)SO_3 – (tentacle type); conditions as in (a) [78]

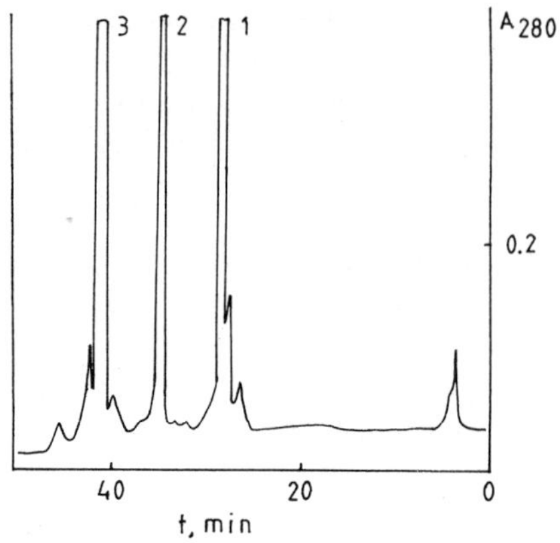

Fig. 12. Fractionation of (*1*) chymotrypsinogen A, (*2*) cytochrome C and (*3*) lysozyme (0.5 mg each) in LiChrospher SO_5^- (5 μm) (tentacle type). Column size, 40 × 10 mm I.D.; flow rate, 1 ml/min. Solvent A = 0.02 mol/l Tris-HCl, pH 8.0; solvent B = A + 1 mol/l NaCl; gradient, 0–100% B in 100 min [78]

Compared with conventional type, the tentacle type exhibits a marked increase in selectivity. The results observed for preparative ion exchangers of tentacle type prompted the author to test this structural arrangement of the binding groups also for analytical materials based on porous silica. Figure 12 reveals that the tentacle-specific selectivity is fully preserved when the matrix is changed.

The change in selectivity is in perfect agreement with the ion exchange mechanism for proteins proposed by Kopaciewicz et al. [106], according to which an analyte will interact with an ion exchange matrix only with a certain, distinct area of its surface. In the tentacle-type exchanger the flexibility of the charge arrangement allows additional electrostatic or other interactions that should account for the changed selectivity.

In a standard ion exchanger the ionic groups are fixed via short arms on the support surface, thus forming a rigid array of binding sites for the poly-ionic analyte. This implies that the analyte may be distorted during the process of maximizing the number of ion pairs formed between the ion exchanger and the analyte. It is obvious that this process may be of special relevance at low ionic strength at which the electrostatic effects are sufficiently large, i.e. conditions under which the analyte is bound. The extent to which such a distortion is reversible or will induce denaturation or irreversible binding is difficult to establish. One can avoid this effect by fixing the ionic groups on linear polyelectrolytes bound to the support surface. This is partially realized in an ion exchanger consisting of silica particles that are coated with ionic polymers and crosslinked in an additional step. In this instance the true mobility of the groups, as provided by the flexibility of the crosslinked polymer chains, can hardly be estimated. A maximum motional freedom of the groups is guaranteed for the ion exchangers described by Mueller [107] because the polymer chains carrying the charges are grafted on the support surface in the absence of crosslinkages.

According to [107] one might expect the following features for the material prepared by the procedure described above:

1) the capacity of the ion exchanger depends exclusively on the surface area of the support;

2) the sorbate is barely in contact with the support surface and is thus prevented from non-specific interactions;
3) owing to the high flexibility of the uncrosslinked polyelectrolyte chains, the charges can easily adopt a configuration that is the optimum for their electrostatic interaction with the sorbate.

Recently, new approaches of sorbent construction for reversed-phase chromatography have been developed. Silicas modified with hydrocarbon chains have been investigated the most and broadly utilized for these aims. Silica-based materials possess sufficient stability only in the pH 2–8 range. Polymeric HPLC sorbents remove these limitations. Tweeten et al. [108] demonstrated the application of stroongly crosslinked styrene-divinylbenzene resins for reversed-phase chromatography of peptides.

Some authors have suggested the use of fluorene polymers for this kind of chromatography. Fluorinated polymers have attracted attention due to their unique adsorption properties. Polytetrafluoroethylene (PTFE) is antiadhesive, thus adsorption of hydrophobic as well as hydrophilic molecules is low. Such adsorbents possess extremely low adsorption activity and nonspecific sorption towards many compounds [109–111]. Fluorene polymers as sorbents were first suggested by Hjerten [112] in 1978 and were tested by desalting and concentration of tRNA [113]. Recently Williams et al. [114] presented a new fluorocarbon sorbent (Poly F Column, Du Pont, USA) for reversed-phase HPLC of peptides and proteins. The sorbent has 20 μm in diameter particles (pore size 30 nm, specific surface area 5 m^2/g) and withstands pressure of eluent up to 135 bar. There is no limitation of pH range, however, low specific area and capacity (1.1 mg tRNA/g) and relatively low limits of working pressure do not allow the use of this sorbent for preparative chromatography.

Silica sorbents covered with fluorene-polymers combine advantages of polymeric as well a silica materials. Saburov et al. [77, 115] described chromatographic sorbents on the basis of controlled pore glasses (CPG) and PTFE. Such sorbents are characterized by extremely low adsorption activity toward almost all classes of compounds. Structural characteristics of the sorbents were examined by means of mercury porometry. Techniques of modification of the silica supports allowing the preservation of its structural characteristics were developed. Composite materials proved to be stable over a broad range of pH (1–14). Organic solvents (ethanol, propanol, acetonitrile, etc.), aqueous salt solution of high ionic strength and 6 mol/l urea did not influence the adsorption properties of the sorbent during elution. In addition, sorbents can be subjected to heat treatment up to 200 °C without any visible changes in chromatographic properties [115].

These sorbents may be used either for selective fixation of biological molecules, which must be isolated and purified, or for selective retention of contaminants. Selective fixation of biopolymers may be easily attained by regulation of eluent polarity on the basis of reversed-phase chromatography methods. Effective isolation of different nucleic acids (RNA, DNA-plasmid) was carried out [115, 116]. Adsorption of nucleosides, nucleotides, tRNA and DNA was investigated. It was shown that nucleosides and nucleotides were reversibly adsorbed on

PTFE-containing silica sorbents from aqueous buffers (for example, 0.01 mol/l tris-HCl, pH 7.0). The adsorption of tRNA was completely suppressed in the presence of small additions of organic eluents (for example, 5% propanol). Proteins were eluted by higher concentrations of organic component (30–70%). The sorbents allowed the development of novel methods for desalting, concentration and deproteinization of tRNA, as well as for the isolation of plasmid from RNA and proteins. 100 ml of 0.02 mol/l Tris-HCl, pH 7.5 + 0.001 mol/l EDTA, pH 8.0 (TE-buffer, pH 7.5) + 0.5 mol/l ammonium sulfate, containing tRNA and proteins, was loaded on the column equilibrated with 0.02 mol/l TE-buffer (pH 7.5). tRNA was quantitatively adsorbed on the column and salts were eluted by the buffer. After entire desalting (the process may be controlled by eluate conductivity) the adsorbed tRNA was eluted by 5 ml of 10% propanol. Thus, deproteinization, desalting and 20-fold concentration of tRNA occurred simultaneously in 40 min. The sorbent capacity to tRNA was 70–75 mg/g. The separation of nucleic acids from proteins is usually carried out by ion-exchange chromatography or by phenolic extraction. The described method of tRNA isolation is an alternative to the existing ones and possesses a number of advantages: high rates of purification and the high capacity of column loading. Experiments have proved PTFE-containing silica sorbents to be of importance for application in gene engineering for isolation of plasmid and other vector DNA by means of reversed-phase adsorption chromatography. The results of chromatography of plasmid pBR 322 separated from proteins and RNA are presented in Fig. 13. RNA, chromosomal DNA and proteins are retained on the column and could be eluted by gradient of acetonitrile concentration (0–50%). RNA is eluted with 10% acetonitrile. It should be emphasized that RNA desorption from the column sorbent was attained by a lower concentration of organic modifier than from a commercially available analog ("Nensorb 20", Du Pont). Moreover, capacity of the sorbents is twice as high and it allows separating not only proteins from DNA (as "Nensorb"), but also DNA from RNA. In the regime of size exclusion chromatography, rapid and

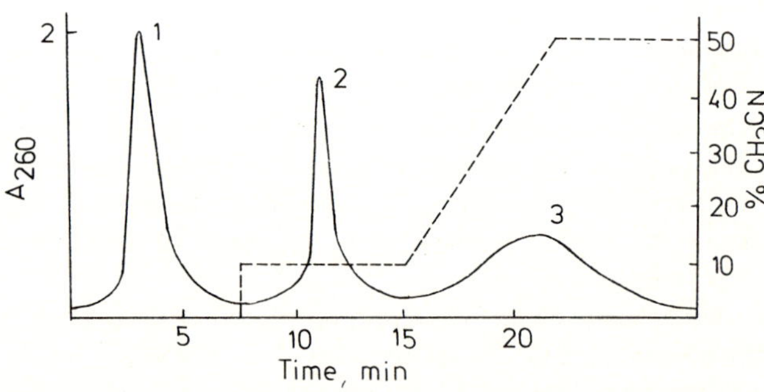

Fig. 13. Separation of (*1*) plasmid pBR 322 (*2*) RNA and (*3*) proteins on controlled pore glass modified with fluorene-polymers. Sample: 5 ml of cleared lysate containing 2 mg of pBR 322. Column size: 10 × 40 mm I.D.; flow rate 1 ml/min. Eluent A = 0.01 mol/l Tris-HCl, pH 7.5; eluent B = A + acetonitrile (1 : 1, v/v), gradient: 0–7.5 min 0% B; 7.5–15 min 20% B; 15–20 min 20–100% B; 20 min ff. 100% B [1]

effective isolation of plasmid (pBR, pUC) and bacteriophage M13 from RNA was carried out [116]. The purity of isolated plasmids was estimated by their full activity as substrates in enzymatic reactions. The CPG-PTFE sorbents possess higher rates of separation (by 8–10 times) in comparison with the commercial sorbent "Sephacryl S-300".

6 Biospecific Chromatography: Advantages of Polymer-Spacered Ligands

Papers on the application of composite sorbents to bioaffinity separations are not very scarce, nevertheless only a few reports of their specific behavior and/or advantages can be found in literature up to now. As we have summarized above, the use of composite sorbents for chromatography of biopolymers involves the positive effects such as diminished non-specific adsorptivity of the stationary phase as well as of its facilitated specific complex formation with solutes. In biospecific chromatography these effects are even more important because the extremely high selectivity of the binding demands improved inertness of the stationary phase and for better adaptivity of the bound ligand. In this section we will attempt to reveal the inherent properties of composite sorbents as compared to the traditional cross-linked polymeric gels and silicas modified by the low molecular weight silanes.

Jervis used porous silica coated with chemisorbed polyacrylhydrazide for immobilization of adenosine monophosphate (AMP) [117]. After periodate oxidation of its ribose residue the ligand was coupled to the carrier and used for isolation of lactate dehydrogenase from rabbit muscle. The specific capacity was 2 mg of protein/g adsorbent with a ligand content of 10 μmol/g, whereas recovery of enzymatic activity after elution was 85%. Hipwell et al. [118] found that for effective binding of lactate dehydrogenases on AMP-ω-aminoalkyl-Sepharose the spacer arm length required at least 4 methylene links. Apparently, a macromolecule of polyacrylhydrazide acts itself like an extended spacer arm and thus allow AMP to bind the enzyme.

Tayot and Tardy [119] have found that the receptor of cholera toxin, so called "ganglioside G_{M1}", attached to dextran-coated silica, was capable of reversible binding to the toxin whereas ganglioside G_{M1} fixed on a cross-linked agarose matrix was not. Indeed, very harsh conditions such as $pH < 3$ and the introduction of denaturants were needed to dissociate the biospecific complex formed between the toxin and ganglioside G_{M1} coupled to cross-linked agarose [120]. Elution of toxin from the composite bioaffinity adsorbent readily occurred in 0.05% citrate-phosphate buffer (pH 2.8) and yielded up to 100% adsorbed toxin as measured by the conventional tests for its determination. Possibly, the affinity of binding was to some extend lowered due to repulsion exerted by the macromolecules of chemisorbed dextran (see Sect. 2.1). Also, multiple hydrogen bonding could contribute to the stronger interaction of the toxin with the condensed agarose-based adsorbent. Therefore, the apparent affinity of a ligand towards solutes could be enhanced or reduced as a result of its coupling to chemisorbed macromolecules. This type of immobilization prevents the direct influence of the surface of the carrier on the association-dissociation processes thus making it more harmonious.

Similar effects were recently observed for sorbent based on polyamide powder with grafted copolymer of hydroxyethyl methacrylate, *N*-vinylpyrrolidone, acrylamide and *N*-dimethylaminoethyl methacrylate explored as a carrier in affinity chromatography [121]. After activation of the prepared carrier with epichlorohydrin, anti-human growth hormone (HGH), anti-human serum albumin (HSA) and anti-insulin antibodies were coupled to epoxy groups. As compared to commercial carriers such as Eupergit C, VA-Epoxy and CNBr-Sepharose, the grafted carrier revealed higher anti-HGH coupling capacity and higher HSA biospecific capacity. The yield of insulin desorbed from the anti-insulin immunoadsorbent was also higher than that of the VA-Epoxy-based immunoadsorbent.

The reason for this difference lies in the unique molecular structure of the grafted carrier. Long tentacle-like spacer arms are introduced into the matrix during grafting, thus rendering the ligands more accessible to high-molecular weight components. There is evidence that the enlarged surface structure rather than the overall porosity plays a decisive role in the separation performance among grafted carriers.

This hypothesis is confirmed by the recent findings dealing with the different biospecific activity of heparin attached to polyvinyl alcohol (PVA) and PEO-hydrogels activated by tresyl chloride [122]. Heparin bound to PEO had an almost ten-fold greater activity than when bound to PVA at comparable concentrations. These observations suggest that the long "leash" provided by PEO hydrogels gives heparin more access to the thrombin-antithrombin pair than the tight bond to PVA, and that crowding of heparin units on a surface limits access by the thrombin-antithrombin pair.

The good function of hydrophilic macromolecular spacers probably arises from their diffuse structure as discussed in Sect. 2.2.

Not only graft but also chemisorbed macromolecules are able to play this role. Strong evidence for this was given by Bootsma et al. [123] who studied the catalytic activity of flavins bound to chloromethylated linear polystyrene in homogeneous, grafted and adsorbed states. In the adsorbed state, a large number of flexible loops and tails of the polymer are extended into the solution, so that the kinetics of aerobic oxidation of 1-benzyl-1,4-dihydronicotinamide was not influenced by the adsorption process. In fact, the kinetic constant k_2 of Michaelis'-Menten's scheme was lower by a factor of 2 for grafted chains than those for adsorbed or dissolved ones. Since no significant desorption of adsorbed polymer was observed, in this case the laborious synthesis of terminally anchored chains offered no special advantages.

Porous glass (PG) modified with covalently adsorbed poly(*p*-nitrophenyl acrylate), as described in Sect. 4.1, turned out to be a highly suitable carrier for immobilization of various biospecific ligands and enzymes. When the residual active ester groups of the carrier were blocked by ethanolamine, the immobilized ligands when bound to the solid support via hydrophilic and flexible poly(2-hydroxyethyl acrylamide). The effective biospecific binding provided by the ligands attached to these chemisorbed macromolecules has been observed when isolating the proteins of interest from sera. Moreover, the non-specific adsorption of other proteins even from highly saturated sera was often negligible. Some examples of separations of this kind are summarized below.

Immobilization of A and B blood group oligosaccharide haptens and preparation of immunoadsorbents with specificity to anti-A and anti-B antibodies has been carried out with the use of polyacrylate-coated PG (WPG-PA) [124]. Prespacered A and B-trisaccharide-β-aminopropylglycosides were used for the synthesis. WPG-PA (1 g) quantitatively binds both haptens (2 μmole) whereas some other activated affinity supports (for example, CNBr-Sepharose 4B) do not. On the other hand, glycidoxypropyl-silica binds prespacered haptens completely but these materials reveal no specific adsorptivity.

The important advantage of the immunosorbents based on WPG-PA is the fast rate of biospecific interaction between the oligosaccharides and antibodies. For B-trisaccharide-WPG-PA the average sorption time of monoclonal B8 antibodies was 20 times shorter than with B-trisaccharide-Sepharose 4B. The role of the flexible polymeric spacer, therefore, is in this case very pronounced.

"Universal" anti-Rh sera deprived of anti-A or B-antibodies were prepared by contacting A and B-immunoadsorbents with human blood sera. To achieve zero titer in the Coombs agglutination test a portion of immunoadsorbent (80–160 mg) proportional to the initial serum titer (1:8–1:64, 1 ml) is required. The incorporation of A-immunoadsorbent into anti-B sera did not interfere with their titer and vice versa. Under the same circumstances an anti-Rh serum titer is lowered by one step or remains unchanged [125]. Properties of this composite sorbent are therefore promising for its use in extracorporal hemisorption processes.

An affinity sorbent based on WPA-PG carrying immobilized human IgG was applied to the isolation of the first component of the complement (C1) from human serum and for its separation into subcomponents C1r, C1s and C1q by the one-step procedure [126, 127]. Cl was quantitatively bound to the sorbent at 0 °C. The activities of subcomponents C1q and $C1r_2r_2$ in the unbound part of the serum were found to be 0.8% and 3.3% of the initial activities in serum. This fraction, therefore, could be used as a R1 reagent for determining the hemolytic activity of C1. Apparently, the neighboring macromolecules of immobilized IgG resemble to some extent an immune complex, whereas C1 formation is facilitated due to the mobility of polymer chains with the attached IgG macromolecules (C1 is usually dissociated in serum by 30%). After activation of bound C1 by heating (30 °C, 40 min) the activated subcomponent C1r is eluted from the sorbent. Stepwise elution with 0.05 mol/l EDTA at pH 7.4 or with 0.05 mol/l EDTA + 1 mol/l NaCl at pH 8.5 results in a selective and quantitative elution of the activated subcomponent C1s and subcomponent C1q.

WPG-PA carrying IgG has been also applied to the isolation of the rheumatoid factor from sera of rheumatoid arthritis patients [128]. Some other applications of WPG-PA are the following.

Peptides formed during tryptic digest of *Salmonella* flagellin were immobilized on the WPG-PG to prepare immunoadsorbents for the isolation of monoreceptor antibodies from rabbit serum against H-antigens of *Salmonella* spp. [129]. The selectivity of the immunoadsorbents coincided with that of the analogous CNBr-Sepharose 4B derivatives.

Heparin was immobilized on the WPG-PA and the sorbent was used for isolation of the fibroblast growth factor from bovine hypophysis. Mitogenic activity of the factor purified on the heparin-WPG-PA and Heparin-Sepharose, as estimated with mice fibroblasts of line N1H-3T3, was virtually the same [130].

Heterogeneous catalysts were prepared by covalent binding of pepsin to wide-porous and nonporous ω-aminoalkyl-derivatized silica compounds, prepared by treatment of polyacrylate-coated silicas with α,ω-dialkylamines. Wide-pore glass-based catalysts showed the maximal retained pepsin activity (92%). The immobilized pepsin content as well as the proteolytic activity of catalyst was found to increase on increasing the chain length of the alkyl spacer from C_2 to C_6, whereas the thermal stability of catalysts (37 °C) was independent of the alkyl spacer length. All the catalysts tested were stable during storage at 8 °C, pH 4.5 for 1 month [131].

The described bioaffinity separations demonstrate that polyacrylamide spacers aid the selective binding of highly complex and delicate biomacromolecules and their associates. Moreover, these solutes remain biologically active after desorption probably due to the high inertness and flexibility of the surrounding polymer chains fixed on the solid support. The unbound parts of serum usually show no loss of the activities of their constituents. Thus we evaluate the surface of inorganic supports coated with chemisorbed *N*-hydroxyethyl polyacrylamide and its derivatives as being biocompatible.

7 References

1. Ivanov AE, Saburov VV, Zubov VP (1989) Zh Vses Khim Ob-va im DI Mendeleeva (in Russian) 34: 368
2. Lee JH, Kopecek J, Andrade JD (1989) J Biomed Mat Res 23: 351
3. Matsunaga T, Ikada Y (1981) J Coll Interface Sci 84(1): 8
4. Ikada Y, Suzuki M, Tamada Y (1984) In: Shalaby SW (ed) Polymers as Biomaterials. Plenum, New York, p 135
5. Absolom DR, Zingg W, Newmann AW (1987) In: Brash JL (ed) Proteins at interfaces. ACS Symposium Series 343, American Chemical Society, Washington, p 401
6. Nagaoka S, Mori Y, Takiuchi H, Yokota K, Tanzawa H, Nishiumi S (1984) In: Shalaby SW (ed) Polymers as biomaterials. Plenum, New York, p 361
7. Lee JH, Kopeckova P, Kopecek J, Andrade JD (1990) Biomaterials 11: 455
8. Klein J, Luckham P (1982) Nature 300: 429
9. Hesselink FTh (1969) J Phys Chem 73: 3488
10. Scheutjens JMHM, Fleer GJ (1979) J Phys Chem 83: 1619
11. de Gennes PG (1981) Macromolecules 14: 1637
12. Scheutjens JMNM, Fleer GJ (1985) Macromolecules 18: 1882
13. Robb ID, Smith R (1974) Eur Polym J (1974) 10: 1005
14. Robb ID, Smith R (1977) Polymer 18: 500
15. Kawaguchi M, Hagakawa M, Takahashi A (1980) Polym J 12: 265
16. Furusawa K, Yamamoto K (1983) Bull Chem Soc Jpn 56: 1958
17. Sakai H, Imamura Y (1987) Bull Chem Soc Jpn 60: 1261
18. Fleer GJ, Lyklema J (1987) Biol Chem Hoppe-Seyler 368: 741
19. Barford W, Ball RC, Nex CCM (1986) J Chem Soc, Faradey Trans 1, 82: 3233

20. Barford W, Ball RC (1987) J Chem Soc, Faradey Trans 1, 83: 2515
21. Luckham PF, Klein J (1984) J Chem Soc, Faradey Trans 1, 80: 865
22. Ball RC, Blunt M, Barford W (1989) J Phys A: Math Gen 22: 2587
23. Douglas JF (1989) Macromolecules 22: 3707
24. Hommel H, Legrand AP, Tougne P, Balard H, Papirer E (1984) Macromolecules 17: 1578
25. Cohen Stuart MA, Waajen FHWH, Cosgrove T, Vincent B, Crowley TL (1984) Macromolecules 17: 1825
26. Peffercorn E, Haouam A, Varoqui R (1989) Macromolecules 22: 2677
27. Cosgrove T (1982) Macromolecules 15: 1290
28. Hiatt CW, Shelokov A, Rosental EJ, Galimore JM (1971) J Chromatogr 56: 362
29. Darling T, Albert J, Russel P, Albert DM, Reid TW (1977) J Chromatogr 131: 383
30. Letot L, Lesec J, Quivoron C (1981) J Liq Chromatogr 4: 1311
31. Dulout C, Peyroset A, Panaris R, Hannoun C, Vincent J (1980) US Patent No 4199450
32. Kopjov VP, Zhigis LS, Reshetov PD, Baburina TM, Brylina VI, Panteleev YuV, Surin VN (1989) Vopr Virusol (in Russian) 6: 760
33. Belousova RV, Lobova TP, Kopjov VP, Ishkildin IB, Zhigis LS, Reshetov PD (1990) Veterenaria (in Russian) 11: 53
34. Tardy M, Tayot JL, Roumyantseff M, Plan R (1978) In: Epton R (ed) Chromatography of synthetic and biological polymers, vol 2, Ellis Horwood, Chichester, p 231
35. Santarelli X, Muller D, Jozefonvicz J (1988) J Chromatogr 443: 55
36. Zhou FL, Muller D, Santarelli X, Jozefonvicz J (1989) J Chromatogr 476: 195
37. Alpert A, Regnier FE (1979) J Chromatogr 185: 375
38. Flashner M, Ramsden H, Crane LJ (1983) Anal. Biochem. 135: 340
39. Lawson TG, Regnier FE, Weith HL (1983) Anal Biochem 133: 85
40. Jastrebov SI, Popov SG (1986) Bioorgan Khim (in Russian) 12: 661
41. Wilchek M (1973) FEBS Lett 33: 70
42. Meiller F, Bonnebat C, Deleuil M (1976) US Patent No 3984349
43. Kurganov A, Kuzmenko O, Davankov VA, Eray B, Unger KK, Trudinger U (1990) J Chromatogr 506: 391
44. Davankov VA, Kurganov AA, Unger KK (1990) J Chromatogr 500: 519
45. Gupta S, Pfannkoch E, Regnier FE (1983) Anal Biochem 128: 196
46. Fausnaugh JL, Pfannkoch F, Gupta S, Regnier FE (1984) Anal Biochem 137: 464
47. Alpert A (1983) J Chromatogr 266: 23
48. Alpert A (1986) J Chromatogr 359: 85
49. Nomura A, Yamada J, Tsunoda KI (1987) Anal Sci 3: 209
49. Gupta S, Pfannkoch E, Regnier FE (1983) Anal Biochem 128: 196
50. Ivanov AE, Zhigis LS, Chekhovskykh EYu, Reshetov PD, Zubov VP (1985) Bioorgan Khim (in Russian) 11: 1527
51. Ivanov AE, Zhigis LS, Turchinsky MF, Kopjov VP, Reshetov PD, Zubov VP, Kastrikina LN, Lonskaya NI (1987) Mol Genet Mikrobiol Virusol (in Russian) 11: 39
52. Nazhmidinov Sh, Turaev AS, Usmanov KhU, Bazarbayev A (1973) J Polym Sci 42: 1591
53. Ivanov AE, Zhigis LS, Reshetov PD, Zubov VP (1987) Proceedings of 3rd All-Union Conference on Water-Soluble Polymers (in Russian) 27–31 May 1987, Irkutsk, USSR, p 196
54. Ivanov AE, Turkin SI, Zubov VP (1985) USSR Patent No. 1,151,295
55. Ivanov AE, Belov SV, Sklovsky MD, Zubov VP (1991) Vysokomol Soed (in Russian) 33B: 289
56. Ivanov AE, Verkhovskaya LV, Khilko SN, Zubov VP (1990) Bioorgan Khim (in Russian) 16: 1028
57. Ivanov AE, Wulfson AN, Jakimov SA, Zubov VP, Arutjunyan AM (1990) Proceedings of 5th All-Union Symposium on Molecular Liquid Chromatography (in Russian). 20–22 November 1990, Riga, USSR, p 172
58. Miller NT, Feisbush B, Karger BL (1984) J Chromatogr 316: 513
59. Chang JE, El Rassi Z, Horvath Cs (1985) J Chromatogr 319: 396
60. El Rassi Z, Horvath Cs (1986) J Liq Chromatogr 9: 3245

61. Miller NT, Feisbush B, Corina K, Powers-Lee S, Karger BL (1985) Anal Biochem 148: 510
62. Papirer E, Nguyen Van Tao (1972) J Polym Sci 10B: 167
63. Iler RC (1979) The chemistry of Silica, Wiley, NY
64. Papirer E, Nguyen Van Tao (1973) Angew Makromol Chem 28: 31
65. Wheals BB (1975) J Chromatogr 107: 402
66. Bryk MT (1981) Polymerization on the solid surface of the inorganic substances, Naukova Dumka, Kiev, USSR
67. Ivanchev SS (1985) Radical polymerization (in Russian), Leningrad, Khimiya
68. Bruk MA, Pavlov SA (1990) Polymerization on solid surfaces (in Russian), Khimiya, Moscow, USSR
69. Babkin IYu, Tzetlin BL (1973) Zh Vses Khim Ob-va im DI Mendeleeva (in Russian) 18: 263
70. Bruk MA (1987) Uspekhi Khimii (in Russian) 56: 148
71. Pavlov SA, Bruk MA, Isaeva GG, Abkin AD (1981) Dokl Akad Nauk USSR (in Russian) 259: 159
72. Kosaka Y, Uemura M, Hashimoto T, Fukano K (1977) US Patent No. 4045353
73. Dinh-Ngoc B, Rabe JG, Schnabel W (1975) Angew Makromol Chem 46: 23
74. Olenin AV, Khristyuk AL, Golubev VB, Zubov VP, Kabanov VA (1983) Vysokomol Soed (in Russian) 25A: 423
75. Kusama Y, Udagawa A, Takehisa M (1979) J Polym Sci Polym Chem Ed 17: 393
76. Bruk MA, Abkin AD, Demidovich VV, Eroshina LV, Urman YaG, Slonim MYa, Ledeneva NV (1975) Vysokomol Soed (in Russian) 17A: 3
77. Saburov VV, Muydinov MR, Guryanov SA, Kataev AD, Turkin SI, Zubov VP (1990) Proceedings of 5th All-Union Symposium on Molecular Liquid Chromatography (in Russian). 20–22 November 1990, Riga, USSR, p 207
78. Mueller W (1988) 8th Intern Symposium on HPLC of Proteins, Peptides and Polynucleotides, Copenhagen, Oct 31–Nov 2, 1988
79. Tutaeva NL, Komarov VS (1975) Dokl Akad Nauk Byelorussian SSR (in Russian) 19: 803
80. Tutaeva NL, Komarov VS, Belyakova MD (1981) Dokl Akad Nauk Byelorussian SSR (in Russian) 25: 45
81. Ivanchev SS, Dmitrenko AV, Shadrina NE, Volkov AM, Ulinskaya NN (1987) Dokl Akad Nauk USSR (in Russian) 297: 402
82. Kalal J, Tlustakova M (1979) Acta Polymerica 30: 40
83. Bleha M, Votavova E, Tlustakova M, Kalal J (1982) Angew Makromol Chem 107: 25
84. Schneider B, Doskocilova D, Stokr J, Tlustakova M, Kalal J (1979) Acta Polymerica 30: 283
85. Gobel G, Starnic J (1978) Angew Makromol Chem 71: 167
86. Ivanchev SS, Golgman AJa, Dmitrenko AV, Krupnik AM, Perepechko II (1985) Dokl Akad Nauk USSR (in Russian) 283: 1225
87. Barabas ES, Grosser F (1976) US Patent No. 3941718
88. Ivanchev SS, Dmitrenko AV (1985) Plast und Kaut (in Russian) 32: 41
89. Korshak VV, Zubakova LB, Kachurina NV, Balashova OB (1979) Vysokomol Soed (in Russian) 21A: 1132
90. Chaimberg M, Parnas R, Cohen Y (1989) J Appl Polym Sci 37: 2921
91. Hayakawa K, Kawase K, Yamakita H (1977) J Appl Polym Sci 21: 2921
92. Mchedlishvili BV, Smirnov AV, Mertvuhina ON, Molodkina LM, Makarova SB, Kolikov VM, Venzel BI, Petuhov VV, Molodkin VM (1984) Colloid Z (in Russian) 46: 132
93. Abuelafiya R, Pesek J (1989) J Liq Chrom 12: 1571
94. Rosevear A, Mattock P (1981) UK Patent No. 1602432
95. Chang SH, Regnier FE (1977) US Patent No. 4029583
96. Gorkovenko AA, Berman EL, Zubov VP, Ponomarenko VA (1987) Izvestiya Akad Nauk USSR, Seriya Khimich (in Russian) 9: 2091
97. Schutijser JA (1983) US Patent No. 4415631

98. Rogers ME, Adlard MW, Saunders G, Holf G (1985) J Chromatogr 326: 163
99. Fukano K, Komiya K, Sasaki H, Hashimoto T (1978) J Chromatogr 166: 47
100. Wehr CT, Abbott SR (1979) J Chromatogr 185: 453
101. Kato Y, Komiya K, Hashimoto T (1982) J Chromatogr 246: 13
102. Mino G, Kaizerman S (1958) J Polym Sci 31: 242
103. Mueller W (1986) Kontakte (Darmstadt) 3: 3
104. Mueller W (1987) Kontakte (Darmstadt) 1: 45
105. Albertsson PA (1971) Partition of cell particles and macromolecules, Almquist Z Wichsell, Stockholm, Sweden
106. Kopacievicz W, Rounds MA, Fausnaugh J, Regnier FE (1983) J Chromatogr 266: 3
107. Mueller W (1990) J Chromatogr 510: 133
108. Tweeten KA, Tweeten TN (1986) J Chromatogr 359: 111
109. Billiet HAH, Schoenmakers PJ, De Galan L (1981) J Chromatogr 218: 443
110. Sadek PC, Carr PW (1984) J Chromatogr 288: 25
111. Xindu G, Carr PW (1983) J Chromatogr 269: 96
112. Hjerten S (1978) J Chromatogr 159: 47
113. Hjerten S, Hellman U (1980) J Chromatogr 202: 391
114. Williams RC, Vasta-Russell JF, Golebowski K (1986) J Chromatogr 371: 63
115. Saburov VV, Turkin SI, Muydinov MR, Ivanov SV, Zubov VP (1986) Proceedings of Chromatography in biology and medicine. International symposium. Moscow, p 200
116. Saburov VV, Vener TI, Gilyarevsky SD, Ivanov SV (1988) 14th International Congress of Biochemistry, Prague, Vol 1, p 235
117. Jervis L (1978) In: Epton R (ed) Chromatography of synthetic and biological polymers, vol 2, Ellis Horwood, Chichester, p 231
118. Hipwell MC, Harvey MJ, Dean PDG (1974) FEBS Lett. 42: 355
119. Tayot JL, Tardy M (1979) French Patent No. 2403098
120. Parikh I, Cuatrecasas P (1974) Methods in Enzymology 34: 610
121. Muller-Schulte D (1990) J Chromatogr 510: 115
122. Tay SW, Merril EW, Salzman EW, Lindon J (1989) Biomaterials 10: 11
123. Bootsma JPC, Wolsink HW, Challa G, Muller F (1984) Polymer 25: 1327
124. Bovin NV, Zemlyanukhina TV, Chagiashvili TsN, Khorlin AJa (1988) Khim Prir Soed (in Russian) 6: 777
125. Chagiashvili TsN, Zotikov EA, Bovin NV, Korchagina EYu (1989) Hematol Transfusiol (in Russian) 8: 56
126. Kozlov LV, Shojbonov BB, Ivanov AE, Zubov VP, Antonov VK (1989) Biokhimia (in Russian) 54: 1745
127. Ivanov AE, Kozlov LV, Shojbonov BB, Zubov VP, Antonov VK (1991) Biomed Chromatogr 5: 90
128. Denisova OV, Myagkova MA, Balabanova RM, Solovyev SK, Ivanov AE, Alyoshkin VA (1988) Lab Delo (in Russian) 9: 48
129. Vaneeva LI, Jankina NF, Ivanov AE, Razgulyaeva OA, Zubov VP (1989) Zh Mikrobiol Epidemiol Immunobiol (in Russian) 5: 50
130. Ivanov AE, Grin MA, Zhiltsov VV, Kizjun SM, Afanasenko GA, Zubov VP (1990) Prikl Biokhim i Mikrobiol (in Russian) 26: 840
131. Ivanov AE, Turková J, Čapka M, Zubov VP (1990) Biocatalysis 3: 235

Editor: Prof. G. Glöckner
Received June 28, 1991

Author Index Volume 101–104

Author Index Vols. 1–100 see Vol. 100

Subject Index